HANDBOOK OF COMMON
NEW GUINEA BEETLES

By J.L. GRESSITT AND R.W. HORNABROOK

WAU ECOLOGY INSTITUTE
HANDBOOK NO. 2
1977

Handbook of Common New Guinea Beetles
Wau Ecology Handbook No. 2
by J. L. Gressitt and R. W. Hornabrook

Originally published by:
  Wau Ecology Institute
  P. O. Box 77
  Wau
  Papua New Guinea
  Tel: 675 + 474 6431

© 1977

ISBN 978-9980-945-16-7

Reprinted with permission by:
  University of Papua New Guinea Press & Bookshop
  P. O. Box 413
  UNIVERSITY PO
  NCD
  Papua New Guinea
  Tel: 675 + 326 7281 or 326 7675
  Email: upngbooks@gmail.com

CONTENTS

ACKNOWLEDGEMENTS

We are very grateful for donation toward publication costs of this book by Dr. Kintaro Baba of Niigata, Japan.

We acknowledge extensive help from the staffs of the British Museum (Natural History), Bishop Museum, Department of Primary Industries in Port Moresby, Forest Research Station in Bulolo, and Wau Ecology Institute. Special gratitude is extended to R. D. Pope, R. T. Thompson, G. A. Samuelson and J. N. L. Stibick. The drawings were mostly prepared by William Adams (including those in color), and partly by Alan Hart (text figures 3, 4, 6, 8, 10, 13, 14, 17, 18, 24, 35, 36); text figure 7 was kindly supplied by M. O. de Lisle. Most of the photographs of Tenebrionidae were loaned by Z. Kaszab; the remaining photographs were taken by J. L. Gressitt, who acknowledges extensive assistance through grants from the U. S. National Science Foundation through Bishop Museum. Other special kindnesses were extended by R. A. Crowson, Z. Kaszab and others. Some identifications were provided by R. J. M. Aldridge, M. E. Bacchus, S. Endrödi, P. M. Hammond, Z. Kaszab, G. Kibby, M. O. de Lisle and W. Wittmer.

THE AUTHORS

J. Linsley Gressitt, PhD. L. A. Bishop Distinguished Chair of Zoology, Bishop Museum, Honolulu; Graduate faculties, University of Hawaii and Papua New Guinea University of Technology; Director, Wau Ecology Institute.

Richard W. Hornabrook, MD, Formerly Director, Institute of Medical Research, Goroka; Professor, University of Papua New Guinea and University of Wellington Medical Faculties; home address: 27 Orchard St., Wadestown, Wellington, New Zealand.

ILLUSTRATIONS
Plates

1. (a) *Therates labiatus* Fabricius
   (Carabidae);
   (c) *Eupatorus beccarii* (Gestro)
   (Scarabaeidae: Dynastinae);
2. (a) *Cladophorus* sp. (Lycidae);

   (c) Bothricara pulchella Guerin
   (Lagriidae);
3. (a) *Coptopterus latus* Gressitt
   (Cerambycidae: Cerambycinae);
   (c) *Glenea lefebueri* (Guerin)
   (Cerambycidae: Lamiinae);

4. (a) *Callistola pulchra* Gressitt
   (Chrysomelidae: Hispinae);
   (c) *Hellerrhinus papuanus*
   (Heller) (Curculionidae:
   Otiorhynchinae);
5. (a) *Dineutes* (*Merodineutes*)
   archboldianus Ochs
   (Gyrinidae);
   (c) Dytiscidae sp.;

   (e) *Plaesius cossyphus* Marseul
   (Histeridae);
   (g) *Priochirus sp.* (Staphylinidae);

   (i) *Phaeochrous emarginatus*
   Castelnau (Scarabaeidae:
   Hybosorinae).
6. (a) *Alaus* sp. (Elateridae);

   (c) *Papuana japenensis* Arrow
   (Scarabaeidae: Dynastinae);
   (e) *Pteroptyx cribellata* (Olivier)
   (Lampyridae);

(b) *Neolamprima adolphinae*
Gestro (Lucanidae);
(d) *Cyphogastra* near *timorensis*
(Buprestidae).
(b) *Encymon bipustulatus*
Gorham (Endomychidae);
(d) *Lagriomorpha indigacea*
Young (Anthicidae).
(b) *Epepeotes rarus* (Thomson)
(Cerambycidae: Lamiinae);

(d) *Promechus pulcher* Gressitt
& Hart (Chrysomelidae:
Chrysomelinae).

(b) *Aphorina australis* (Heller)
(Apionidae);
(d) *Eupholus geoffroyi* Guerin
(Curculionidae: Leptopiinae).

(b) *Hydaticus pacificus* Aube
(Dytiscidae);
(d) *Hydropholus picicornis*
Chevrolat (Hydrophilidae);
(f) *Diamesus osculans* Vigors
(Silphidae);
(h) *Creophilus albertisi* Fauvel
(Staphylinidae);

(b) *Dermolepida noxium*
Britton (Scarabaeidae:
Melolonthinae);
(d) Nitidulidae sp.;

(f) *Luciola* near *obsoleta* Olivier
(Lampyridae);

v

(g) *Chauliognathus gibbosus* Wittmer (Cantharidae);

(i) *Cladophorus?* sp. (Lycidae).

(h) *Cladophorus* sp. (Lycidae);

7. (a) *Hectarthrum trigeminum* Newman (Cucujidae: Passandrinae);

(b) *Carphurus viridipennis* Wittmer (Melyridae);

(c) *Simocoptengis* sp. (Erotylidae);

(d) *Anisolema* sp. (Coccinellidae);

(e) *Neda fuerschi* Bielawski (Coccinellidae);

(f) *Epilachna* sp. (Coccinellidae);

(g) *Dryptops phytophorus* Samuelson (Colydiidae);

(h) *Bradymerus nigerrimus* Gebien (Tenebrionidae: Boleophagini);

(i) *Achthosus auriculatus* Gebien (Tenebrionidae: Ulomini).

8. (a) *Setenis illaesicollis* Fairmaire (Tenebrionidae: Tenebrionini);

(b) *Euhelaeus speculiferus* Gebien (Tenebrionidae: Helaeini);

(c) *Encara devidiens* Gebien (Tenebrionidae: Helaeini);

(d) *Tabarus infernalis* Gebien (Tenebrionidae: Eutelini);

(e) *Pelecotomoides* sp. (Rhipiphoridae);

(f) *Sphingnotus insignis* Perroud (Cerambycidae: Lamiinae);

(g) *Agrianome loriae* Gestro (Cerambycidae: Prioninae);

(h) *Potemnemus* near *detzeri* Kriesche (Cerambycidae: Lamiinae);

(i) *Batocera wallacei* Pascoe (Cerambycidae: Lamiinae).

9. (a) *Promechus pittospori* Gressitt & Hart (Chrysomelidae: Chrysomelinae);

(b) *Promechus paniae* Gress & Hart, larva, dorsal view;

(c) Ditto, lateral view (Chrysomelidae: Chrysomelinae);

(d) *Gronovius imperialis* Jacoby (Chrysomelidae: Galerucinae);

(e) *Ancylotrichis waterhousei* Jekel (Anthribidae);

(f) *Aspidomorpha australasiae* Spaeth (Chrysomelidae: Cassidinae);

(g) *Hellerhinus* sp. (Curculionidae: Otiorhynchinae);

(h) *Gymnopholus nodosus* Gressitt (Curculionidae: Leptopiinae).

10. All Curculionidae:

(a) *Pantorhytes lichenifer* Gressitt (Brachyderinae) with 2 kinds of lichen growing on elytra: *Usnea* above and *Parmelia* below;

(b) *Vanapa oberthueri* Pouillaude (Hylobiinae);

(c) *Rhinoscapha richteri* Faust (?) (Leptopiinae);

(d) *Eupholus bennetti* Gestro (Leptopiinae);

(e) *Carbonomassula cobaltina* Heller (Molytinae);

(f) *Rhyncophorus bilineatus* Montrouzier (Rhyncophorinae);

(g) *Disopirhinus* sp. (Cryptorhynchinae);

(h) *Arachnopus* sp. (Zygopinae).

TEXT-FIGURES

1. Beetle anatomy, dorsal view.
2. Beetle anatomy, ventral view.
3. *Colpodes habilis* Sloane (Carabidae: Agonini).
4. *Tricondyla aptera* Olivier (Carabidae: Cicindelinae).
5. *Hesperus abnormis* Cameron (Staphyliniae).
6. *Cyclommatus sumptuosus* Moellenkamp (Lucanidae).
7. *Aegus gressitti* de Lisle (Lucanidae), male left, female right.
8. *Pelopides schraderi* Kuwert (Passalidae).
9. *Onthophagus latinasutus* Arrow (Scarabaeidae: Coprinae).
10. *Lomaptera lutea* (Linnaeus) (Scarabaeidae: Cetoniinae).
11. Family Helodidae.
12. *Callirhipis* sp. (Callirhipidae).
13. *Cyphogastra timorensis albertisi* Gestro (Buprestidae).
14. *Alaus* sp. (Elateridae).
15. *Luciola obsoleta* Olivier (Lampyridae)
16. *Attagenus undulatus* (Motschulsky) (Dermestidae).
17. *Xylothrips religiosus* (Boisduval) (Bostrychidae).
18. *Omadius semicarinatus* (?) (Cleridae).
19. Family Melyridae.
20. ?*Carpophilus* sp. (Nitidulidae).
21. *Mimemodes* sp. (Rhizophagidae).
22. ?*Heliota* sp. (Cucujidae).
23. *Coenolanguria papuensis* (Crotch) (Languriidae).
24. *Epilachna haemorrhoa* Boisduval (Coccinellidae).
25. *Lyprops atronitens* Fairmaire (Tenebrionidae: Lypropini).
26. *Glipa mixta* Fabricius (Mordellidae).
27. *Macrosiagon cucullatus* Macleay (Rhipiphoridae).
28. *Ananca kanak* Fairmaire (Oedemeridae).
29. Family Anthicidae.
30. *Tmesisternus flavolineatus* Breuning (Cerambycidae: Lamiinae).
31. *Acanthoscelides obtectus* (Say) (Bruchidae).
32. *Aulacophora pallidefasciata* Jacoby (Chrysomelidae: Galerucinae).
33. ?*Ithystenus* sp. (Brenthidae).
34. *Behrensiellus glabratus* Pascoe (Curculionidae: Otiorhynchinae).
35. *Xyleborus sordicavola* Schedl (Curculionidae: Scolytinae).
36. *Crossotarsus bironeanus* Schedl (Curculionidae: Platypodinae).

INTRODUCTION

The beetles, including the weevils, make up the order Coleoptera, which may be the largest order of insects. The Hymenoptera (wasps, bees and ants) and the Diptera (true flies) include numerous species, many of them very small, but the beetles are better known, with more named species. Beetles range in body size more than insects of most other orders, and they are found almost everywhere on land.

Although this handbook is entitled "Common New Guinea beetles" it includes only selected representatives. Probably more than 25,000 species of beetles occur in New Guinea and nearby islands. Many hundreds of these are certainly abundant in one part of the island or another, and many occur widely in the lower altitude areas. Most are not well known scientifically and are difficult to identify. Many are still without names. The order is so large that many specialists would have to work for years before the group could become fairly well known here. This small handbook gives only the simplest introduction to the subject.

There are well over 100 families of beetles (Crowson, 1967), most of which occur in New Guinea. Just over one-third of these have been selected for this handbook, mainly the larger and more conspicuous groups. They include about 90% of the kinds of beetles that the average student might find in New Guinea without using special methods for the very small species. Because there are so many families of beetles, and so very many species, it is not easy to provide ready means of identification. The student will have to supplement the use of the incomplete key provided below with the photographs and drawings, together with the notes on the various families. References are also provided for further help.

It must be borne in mind that there is not space to include all known facts concerning New Guinea beetles, and more is omitted than is included. The help of specialists is needed to identify the bulk of the species. This is mainly a guide to the families, and the authors are more familiar with the leaf-beetles, longicorn beetles and some groups of weevils.

Much of the early pertinent work on New Guinea beetles was the result of studies made on collections from expeditions by Germans, British, Hungarian, French and Italians. Only more recently have Australian, American, Japanese, New Zealand, Indonesian and Papua New Guinean workers become involved.

Wau Ecology Institute, the Entomology (Primary Industries) Laboratory at Konedobu (Port Moresby), the Forest Research Station at Bulolo, the Bishop Museum in Honolulu and the British Museum (Nat. Hist.) in London have extensive identified collections of New Guinea beetles. Some may also be seen at the National Collection (C.S.I.R.O.) in Canberra, at the Australian Museum in Sydney and the South Australian Museum in Adelaide, as well as at some major museums in Europe and America.

New Guinea as used in this work refers to the entire island of New Guinea plus nearby islands (Irian Jaya plus Papua New Guinea).

THE PLACE OF BEETLES IN THE INSECT WORLD

Beetles may amount to over 30% of the total number of kinds of insects. Like all true insects, they have three pairs of legs and a pair of antennae, but differ from nearly all others in having the fore pair of wings (elytra) hardened to help form a more or less rigid case covering the hind-thorax, the abdomen and the pair of membranous hind wings which are used for flying. Some beetles have short elytra which do not cover all, or much, of the abdomen; in these the hind wings are usually folded under the elytra when not in use, and the top of the abdomen may be fairly tough (well sclerotized). Earwigs (Demaptera) may fit this description, but have the abdomen ending with a forceps-like pair of cerci. Some beetles cannot fly and may have the hind wings and their wing-muscles more or less reduced. Some of these may even have the elytra fused.

There are about 250,000 known species of beetles in the world but many are still unnamed. They include some of the largest insects, and many are of striking pattern or form. Beetles also play many important roles. In many situations they are the most abundant insects.

OCCURRENCE

There are beetles in nearly all terrestrial and fresh-water environments, in most situations where insects can survive, and even some in the inter-tidal zone on beaches or in mangrove and other brackish swamps. Some live as high as the summit of Mt Wilhelm. The fresh-water beetles belong to several families and include both predators and scavengers. These mostly carry bubbles of air with them, usually under the wing-covers or beneath the body and held by hairs. The larvae of many of these have gills for taking oxygen from the water so do not have to come to the surface for air.

While beetles occur in nearly all situations, only a few are parasitic, being found in the fur of certain mammals.

In tropical areas, such as the lowlands of New Guinea, there are very many species, with usually fewer individuals of one kind together; but in more temperate climate, as in high mountains, species tend to be fewer, but individuals of one kind may be relatively abundant. Some species are very widely distributed; others are of limited natural range. The former are more characteristic of lowland New Guinea, and the latter of the highlands. For instance, one species may be restricted to one mountain range or valley, and often within a limited altitude range.

To indicate altitude range of species or genera in the text, we use the four beetle species—zones proposed for Papua New Guinea by Dr Szent-Ivany in 1966:

I. Coastal and lowland, 0–450 meters, rarely to 900 meters altitude
II. Wider ranging forms, from 0–2200 meters altitude
III. Mid-montane, 1400–2500 meters altitude
IV. Upper montane, 1800–3600 meters (also alpine to 4500 m.)

SIZE AND DIVERSITY

Beetles range in size from a fraction of a millimeter to more than 200 mm in length, with the largest in New Guinea about 125 mm long.

Beetles vary greatly in body structure. The ground beetles have a so-called primitive or generalized body form, with somewhat flattened oval body and fairly long slender legs and antennae. Their close relatives the tiger beetles have still longer legs and are very fast on the wing. The water beetles are mostly flattened boat-shaped and have special adaptations for locomotion in water and for breathing. The rove beetles have very short elytra and many are active flyers. The solder-beetles, fire-flies and net-winged beeltes have rather soft elytra, unlike most other beetles. Click-beetles jump by means of a hinge-like joint in the thorax. Their relatives the buprestids can take flight very quickly but have short legs. The clerid beetles are also quick to fly and are usually seen on fallen trees. Nitidulids are small and flattened, with shortened elytra. The coccinellids or lady-bird beetles are round or oval and convex. The usually stout lamellicorns include the scarabs and their relatives, many of which have strong spiny legs and burrow in soil, dung or fleshy or rotten plants. The males of some have conspicuous horns and the male stag beetles often have very long jaws. The chrysomelids are leaf-feeders and of small to medium size. The bruchids are short-bodied and mostly feed in peas and beans. The long-horned beetles are mostly slender, with long antennae, and some are very large. Tenebrionids are diverse, with long or short legs, the former mostly in trees and the latter under bark or on the ground. Members of many other families live under dead bark and are greatly flattened, or quite small. Lagriids have a narrow prothorax and wide elytra and occur on trees and logs. Colydiids are of various shapes and are usually associated with wood.

The weevils are very numerous in species, and vary widely in size and form. Their jaws are on the end of a long or short snout. The anthribids may have very long antennae and can be confused with cerambycids; they are also wood-borers, or live in seeds. The brenthids are long and slender and are also wood-borers. The bark-beetles are much smaller, and more or less cylindrical and live under bark, or in stems.

BIOLOGY AND LIFE HISTORY

Beetles have complete metamorphosis, meaning with four distinct stages in the life cycle: egg, larva, pupa and adult. Reproduction is almost always bisexual. Males are often smaller than females and may have longer antennae. The male reproductive organ consists partly of a rigid tube which can be extended from the tip of the abdomen to enter the bursa copulatrix in the tip of the female's abdomen. The sperm are stored in the spermatheca in the female abdomen until needed to fertilize the eggs.

The eggs are usually deposited singly or in clusters on or near the larval food, in protected places, such as under bark, in stems, carrion, dung, etc, but some are glued to leaf surfaces. The eggs may be naked or enclosed in glue-like or wax-like material or in a papery sheath.

The larva, which hatches from an egg, feeds until a limit is reached, when its skin is stretched. It then rests a while and sheds its skin. This is called molting, and occurs a definite number of times (usually 3–5) depending on the group. The mature last stage (or instar = stage separated by molts) larva prepares a place for pupation.

The pupa is formed by shedding the skin of the last instar larva after a resting period. During the pupal stage the organs are transformed from those of the larva to those of the adult beetle. The pupal stage may take place in a cell in ground or wood, or in a leaf-mine, or in some cases the pupa may simply be attached to a leaf surface or lie free under loose bark. The pupa is essentially immobile and defenceless and is the most vulnerable stage, subject to attack by predators and parasites.

The adult emerges by shedding the pupal skin, and takes some hours or days to harden and assume adult coloration. When ready it chews its way through the bark or works its way out of its cell.

Beetle larvae do not resemble adults of the species, being specially adapted to their environment. They are usually out of sight, and most are white or pale yellow, but some living partly exposed in humus, dung or rotting logs may be black or dark brown and have tougher skin. This is true of most of the predatory larvae. Water beetle larvae are usually long and tapered, but some are flat and oval. They may have external gills or air tubes. In New Guinea certain leaf-beetle larvae, living between the bases of palm fronds where water collects, have the same flat-oval form. Most leaf-beetle larvae are pale to brown, soft-bodied and naked or with bristles or processes. Sometimes the bodies, or the molted skins and bristles, are covered with the larva's excrement for protection and camouflage. Larvae of click-beetles are usually very slender, cylindrical and hardened, some feeding on roots and some in rotten wood as predators. Some larvae which scavenge under rotting bark (tenebrionids or others) are also of this form. Lady-bird beetle larvae may be brightly colored, with spines or waxy secretion. Those of boring beetles are mostly pale, with dark head, soft-bodied and elongate. They may be cylindrical, ovate or flattened. Those of lamellicorns (scarabs and relatives) are strongly arched, so that the head and true legs are close to the end of the abdomen. Many weevil larvae are somewhat similar in shape, but usually have the hind portion of abdomen more swollen, and are less hairy. A few beetles have more than one type of larva, with change in feeding habits, usually after the first instar. This kind of development is called hypermetamorphosis, and occurs in Rhipiphoridae and Meloidae.

As beetle larvae occur in many forms, a simple definition for their recognition is difficult. However, very few can be confused with the caterpillars of moths and butterflies, or with the minute-headed white larvae of most bees, wasps and ants, with the apparently headless maggots of most flies, or with the nymphs of bugs, grasshoppers, etc, which are similar to their respective adults but without wings. Time-span of the larval period may be from a few weeks to several months.

The pupal stage of beetles is a resting period for the transformation (metamorphosis) from larva to adult, as with the chrysalis of a butterfly or moth. Usually the pupa is naked like the larva; rarely it may be enclosed in a cocoon-like case. In most cases the form of the pupa is quite suggestive of the adult except for the pale color and the unexpanded form of the wings. The pupal stage may last from a few days to a few weeks. When the pupa changes to adult, the appendages are freed, wing-covers and wings quickly expand to full size and pigmentation develops. The adult is termed teneral during this period before the exoskeleton is fully hardened and pigmented.

The adult life-span generally lasts for a few weeks to a few months, but adults of some *Gymnopholus* weevils in New Guinea mountains live for five years or more. Thus the total life-cycle of a beetle may last from a few weeks to several years, but a period of 2—6 months is probably most common for those in lowland New Guinea. Some species can thus complete five or six generations per year. A female may lay eggs in moderate numbers at intervals over a long period. In general, beetles tend to live longer than other insects.

ECOLOGY AND ECONOMIC IMPORTANCE

Beetles play important roles in the interactions, such as food-chains, in the various ecosystems. Since beetles form a considerable fraction of insect species and individuals, and since they are larger in average size than most insects, they make up an important segment of the biota and of the biomass.

Beetles have harder exoskeletons than most insects, and their larvae are usually in protected environments, so they may not proportionately form as important a component of the food of birds as do some other insects. On the other hand, some other insects have developed distastefulness to birds and some other predators. Some mammals eat beetles; lizards, and especially skinks, are important beetle predators, as they search on plants, logs and in the upper humus layer. Frogs and toads are important too. Most beetles have parasites, which may be wasp or fly larvae, or lower animals such as nematode worms or certain protozoans. The parasites

5

usually develop in beetle larvae, but some flies parasitize adult beetles. Predatory insects feed partly on beetles.

As predators, beetles play a very great role. The carabids, including tiger beetles, as well as most of the groups of water beetles, most staphylinids, and some others living under bark or in moss, are predaceous. These help keep down the numbers of many types of insects and other invertebrate animals, such as land snails. The lady-bird beetles (with a few exceptions) are very effective in controlling very important crop pests among aphids, white-flies and scale insects.

As plant-feeders, beetles are of great importance, attacking all parts of plants. Leaf-feeders (especially chrysomelids, weevils, some scarabaeids) can defoliate plants, although usually they are less significant than caterpillars. Some chrysomelids and buprestids are leaf-miners. Other beetles bore under bark, or in heart wood (buprestids, cerambycids), feed on buds, flowers, fruit (weevils, scarabaeids, etc). Many trees are killed by beetles of one species or by several kinds in combination. Some bark beetles carry spores of fungal diseases of trees. The seeds of many trees are destroyed by members of several families of beetles. Larvae of many chrysomelids, scarabaeids and weevils are root-feeders, while the adults feed on leaves, buds and flowers.

Many groups of beetles are scavengers and perform the useful functions of breakdown of dead plant materials and formation of humus and soil. These processes involve various boring beetles (cerambycids, weevils, bark beetles) breaking down the dead wood and many others feeding in the decaying wood and other vegetable debris. Those which feed in grain and other stored plant products such as drugs, lumber, furniture, etc, are pests. But dung feeders disperse dung in the ground while provisioning nests for their larvae, and in so doing reduce the breeding of pest flies and aerate and fertilize the soil. The carrion beetles mostly perform these same beneficial functions in helping to break down the corpses of animals. Some dermestids do of course damage dried meat, hides, other animal products and animal collections, as in museums. But certain of these are actually used in museums for cleaning meat off of skeletons.

Beetles of certain families (especially some staphylinids and paussids) have become adapted to living in nests of termites or ants. In general there is some mutualistic or symbiotic relationship and the beetles may have become greatly modified in form so that they are dependent upon their hosts for food; but in turn provide special food for their hosts from glands apparently developed for this purpose.

Some beetles are able to produce light (fire-flies and a few others), and are almost unique in this respect among insects or terrestrial animals.

BIOGEOGRAPHY

Biogeography is the study of the occurrence of species (plants and animals)

6

throughout the world, on land and in sea. This means not just the geographical range of groups and species, but altitudinal, environmental and ecological occurrence. Thus the field overlaps most aspects of biology and ecology, and requires insight into all species' habits, movements, feeding, breeding, niches and inter-relationships. It is a complex subject and has important application to agriculture, forestry, public health, and other material or basic needs of man, as well as to conservation of natural resources. The complex ramifications cannot be fully dealth with; we shall only touch on this subject and suggest possible origins and relationships of some of the elements of the beetle fauna of New Guinea.

From the standpoints of plants and of insects, many biogeographers will agree that much of the biota of New Guinea is Oriental — that is, originating from the areas to the west of New Guinea, including SE Asia, especially Indonesia and the Philippines. On the other hand, to a vertebrate zoologist, and according to most textbooks, New Guinea is part of the Australian Region, because the mammals in particular, and the birds and reptiles to a considerable extent, are related more to the Australian fauna. In explanation, it can be assumed that there has never been continuous land connections with the islands west of New Guinea, whereas New Guinea and Australia were connected by land during several periods, at least in the Pliocene and late Pleistocene. For one thing the ice ages in the latter meant that the sea level was far lower, exposing the shallow sea bottom between New Guinea and Australia; thus land animals were able to move between these two land-masses. But insects and most plants are adept at crossing narrow sea barriers, which explains why they are the ones which came from the west; besides, climate of New Guinea was amenable to the forms of life in those islands. The relationships with Australia are likewise to a great extent ecological: many of the Australian elements in New Guinea are limited to the open savanna areas of the south, and New Guinea elements in Australia are largely limited to the northern Queensland rainforests. The lack of invasion of placental mammals other than some bats and rodents from the west left vacant niches into which marsupials radiated. Most beetles reached New Guinea from the islands to its west, and apparently far fewer from Australia. These invasions took place over a very long period of time, say a few million years, and during this period various groups have evolved genera now unique (endemic) to New Guinea, or characteristic of New Guinea and nearby areas. There are clear relationships also among beetles to those in New Caledonia and New Zealand. Possible relationships with groups in southern South America have not been well worked out, but might parallel the "trans-antarctic flora" (*Araucaria, Nothofagus*, etc) resulting from the spread of the southern continents after the break-up of Gondwanaland in the Mesozoic Era.

The New Britain fauna has distinct differences from that of mainland

New Guinea, as well as essentially lacking the montane elements. Bougainville fauna is definitely Solomons in nature, differing much further, with some impoverishment. New Ireland's fauna is somewhat intermediate between those of New Britain and the Solomons, but is closer to New Britain's. Bougainville's is much closer to that of Guadalcanal than to New Britain's or New Ireland's. (See Gressitt, 1961, 1974).

BODY STRUCTURES

Adult beetles have most of the essential structures common to other adult winged insects. However, because the elytra form a shell over the hind wings and hind body, they are quite different in appearance from Lepidoptera (moths and butterflies), Diptera (true flies: 1 pair of wings only), Hymenoptera (ants, bees and wasps, with usually 2 pairs of membranous wings), and Neuroptera (lace-wings, ant-lions). The latter may be the closest relatives of the beetles. Some beetles much more nearly resemble cockroaches superficially; but the latter are much more primitive than beetles and have incomplete metamorphosis (egg, nymph, adult instead of

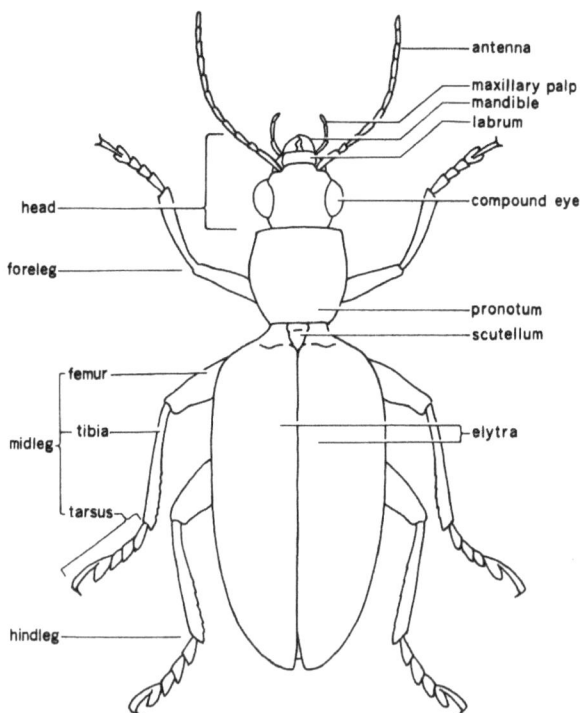

Figure 1 Beetle anatomy, dorsal view.

8

egg, larva, pupa, adult), contrary to lace-wings and ant-lions.

The pincipal body portions are head, thorax and abdomen. The head bears chewing mouthparts with usually strong mandibles, with clypeus and labrum above and maxillae and labium beneath. Labrum and labium may be likened to upper and lower lips, but the two mandibles are at each side of the mouth and move sideways. Maxillae and labium each bear a pair of palpi or tasters. Behind the clypeus is the pair of antennae, which may be simple, clubbed, lamellate, elbowed or toothed, and often have 11 segments, or more or fewer. Below or behind the antennal insertions is the pair of compound eyes. Most beetle lack the ocelli or simple eyes, of which most insects have two or three near the upper edges of the compound eyes. The neck is inserted into the anterior opening of the prothorax, and sometimes the anterior margin of the pronotum hides part or most of the head.

The thorax has the usual three divisions, pro-, meso- and metathorax, each bearing one of the three pairs of legs, and the meso- and metathorax bear the two pairs of wings. However, in beetles the prothorax is less firm-

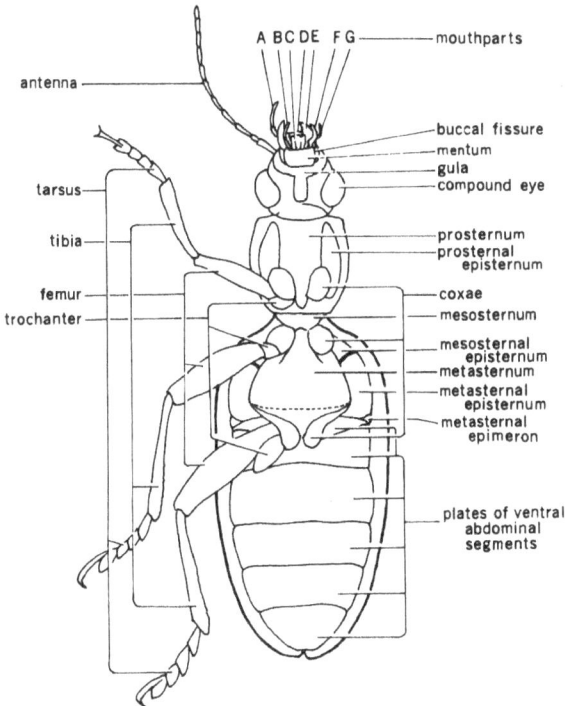

Figure 2 Beetle anatomy, ventral view.

9

ly attached to the mesothorax than in many insects, and from above most beetles appear to consist of three portions which are head, prothorax and elytra. The latter, and the scutellum which is often visible between their bases, belong to the mesothorax, and the elytra usually hide this plus the metathorax, the hind wings and most or all of the abdomen. The meta-thorax and abdomen are usually rather broadly and not very flexibly attached. The legs each consist of femur, tibia and tarsus, the latter of 2—5 segments plus usually paired claws; also coxa and trochanter which provide the basal attachment and articulation.

At each side of most of the abdominal segments, often hidden by the elytra, is a spiracle, an opening of the tracheal or air-tube system of respiration, and there is usually a spiracle also on each side between the meso-and metathorax.

The internal organs include the alimentary canal, shorter in predators and longer in plant-eating and scavenging types. The circulatory system is simple with a partly open dorsal vessel (pumping heart, or aorta) and the blood, rarely red and often greenish, flowing through the body cavity. The nervous system consists of the paired ventral nerve cord with swellings (ganglia) giving off branches in each segment, the head ganglion being larger than the others, with nerves leading to the eyes, antennae, palpi and other sensory organs. The muscular system is mainly for movement of the wings, legs and other appendages and thus is concentrated in the thorax.

Larval structures may appear to be quite different from those of the adults, because of the usual rather different body form. However, they are essentially the same in origin, if not in obvious function, and change from one to the other through metamorphosis. The abdomen is usually relative-ly longer in the larva than in the adult. The larva lacks externally visible wing-pads which are characteristic of most immature insects with incom-plete metamorphosis, and also lacks obviously developed reproductive organs. Larvae may have well-developed legs, minute legs or no visible legs. Those in the two latter categories may crawl by means of swellings on thoracic and abdominal segments, as in many wood-boring species.

Various beetles have structural adaptations for protection, such as horns, large mandibles, glands producing smelly secretions, or others. Some ground beetles emit a fluid which produces a small explosion with brief high heat locally. This can burn and stain human skin. Blister beetles and oedemerids also emit fluids which burn tender human skin. Some beetles produce special sounds to frighten enemies, by rubbing two struc-tures together (stridulation), such as hind leg and elytral margin, or pro-and mesothorax. Some beetles nurture plant growth, or place mud or excrement on their backs for camouflage. Bark beetles have modifications for pushing out debris from their tunnels. Many beetles, such as cryp-torhynchine weevils, feign death for protection, folding in their appendages closely so that they resemble seeds or pebbles.

The beetles are currently assigned to four suborders (one not treated in this book), and many superfamilies. These are mentioned in the key, and also as headings in the main text. With over 100 families of Coleoptera, it is difficult to present a ready means of identification. Some basic characters used in classification are very difficult to observe without dissection, making the use of a key often rather frustrating. It is simpler to learn general recognition points, involving a combination of body proportions, size, shape and contours, together with type of antenna, tarsi or other points. However, the latter system will lead to some errors, so it is desirable to use basic characters as far as possible.

There have been many classifications for beetles, and ideas on this subject have evolved considerably over the years. Thus different textbooks may present differing classifications. The one used in this book is patterned largely after Crowson, 1967. (See *Systematic treatment,* page 17).

Scientific names usually consist of three parts: Genus (first letter always capitalized); species (first letter always small); and author of the species (published first in a scientific journal). The author's name is in parentheses if the species was later transferred to a different genus.

Superfamily names end in -oidea; family names end in -idae; subfamily names end in -inae; and tribal names end in -ini.

KEY TO COMMON NEW GUINEA FAMILIES OF BEETLES

This key is partially adapted from Crowson, 1954 (reprint 1967). It can not be expected to properly function on every use. Not only are many families omitted, but characters are greatly simplified, so many cases may occur where specimens do not run properly through the key. The attempt has been made to use simpler characters, omitting difficult ones and those based on internal anatomy, which may be more fundamental. Superfamilies, suborders and other higher categories are in parentheses. Some larval characters are used to supplement other points. Some families occur at more than one step in the key because of inconsistent characters.

1. Hind coxae rarely fused to metasternum, not dividing abdominal segment 1; abdomen rarely with 6 visible segments 2
   Hind coxae fused to metasternum, dividing abdominal segment 1; abdomen with 6 visible segments; legs of larvae with distinct tarsus and claws (suborder Adephaga) 3
2. Metepisternum reaching mid coxal cavity; hind wing spirally rolled when at rest (suborder Archostemata); body slender, parallel-sided, flat above; antenna slender (page 19) . . . . .Cupedidae
   Metepisternum rarely reaching mid coxal cavity; hind wing not spirally rolled when at rest (suborder Polyphaga) 6

3. Hind coxa not meeting elytral border when at rest; antenna partly pubescent 4

Hind coxa meeting elytral border at rest; antenna entirely glabrous; all species living in water. 5

4. Metasternum without transverse suture in front of hind coxa; median portion of 1st visible abdominal sternite exposed between coxae (p. 19). . . . . .Rhysodidae

Metasternum with transverse suture in front of hind coxa; median portion of 1st visible abdominal sternite not exposed between coxae (p. 19). . . . . .Carabidae

5. Eye completely divided; antenna short and thick, segment 2 large; fore leg longest, middle and hind legs short, paddle-like larvae with lateral gills (p. 24) . . . . Gyrinidae

Eye not divided; antenna slender, segment 2 not large; fore leg short, mid and hind legs longer, often fringed; larva without lateral gills (p. 23). . . . . .Dytiscidae

6. Antenna often with a club involving last 5 segments; tarsi usually 5-segmented; larva often with articulated cerci 7

Antenna rarely with a club involving last 5 segments; tarsi often with apparently 3 or 4 segments, or 5-5-4; larva without articulated cerci 15

7. Antenna with last 3−7 segments never forming a club of plates; body usually not very stout; larva with cerci. 8

Antenna with last 3−7 segments forming a club of plates extending out on one side; body stout; larvae lacking cerci (Lamellicornia or Scarabaeoidea) 11

8. Antenna short, segments 7−11 forming a pubescent club; maxillary palp usually longer than antenna; head with Y-shaped groove between eyes; mostly water beetles (Hydrophiloidea) (p. 24) . . . . Hydrophilidae

Antenna simple, longer than maxillary palp; head lacking Y-shaped groove; not water beetles 9

9. Antenna usually simple (not elbowed or clubbed); elytra often abbreviated (Staphylinoidea). 10

Antenna elbowed, with last 3 segments forming a club; elytra truncate, not greatly abbreviated (Histeroidea) (p. 25). . . . . .Histeridae

10. Elytron very short, leaving several segments uncovered; body usually quite slender (p. 25). . . . . Staphylinidae

Elytron entire or slightly abbreviated; body broad (p. 25). . Silphidae

11. Abdomen with 6 visible sternites, rarely sharply set off from vertical pleurites; abdominal spiracles lateral 12

Abdomen with 5 visible sternites, usually flat and sharply set off from pleurites; abdominal spiracles dorsal, covered by elytra. 13

12. Antenna 8−10 segmented; hind tibia with 1 apical spur (p. 30) . . . . Scarabaeidae

Antenna 11-segmented; hind tibia with 2 apical spurs (p. 35) . . . . Geotrupidae

13. More than 3 plates in antennal club, the plates not closely fitting; species long and flattened 14
    Antennal club with 3 closely fitting plates; species short, subrounded and convex (p. 35) . . . . Trogidae
14. Antenna elbowed, club segments not capable of fitting together (p. 27). . . . . Lucanidae
    Antenna not elbowed, club segments capable of loosely touching (p. 29). . . . . Passalidae
15. Hind coxa excavated posteriorly to receive retracted femur; fore coxal cavity open externally and internally; antenna rarely clubbed . . 16
    Hind coxa rarely excavated to receive femur; fore coxal cavity closed or partly closed externally or internally; antenna frequently clubbed 24
16. If fore coxa projecting, labrum not distinct or radial cell of wing long 17
    Front coxa projecting; labrum free; abdomen with 5 visible sternites (Eucinetoidea); body quite round, convex, smooth; tarsus with segment 4 lobed below; head with keels below eyes (p. 35) . . . . Helodidae
17. If fore coxa transverse, antenna usually more or less serrate; tarsi with segments 3 or 4 usually lobed; last tarsal segment rarely as long as preceding 3 together 18
    Fore coxa transverse to slightly projecting; if rounded, hind coxa without distinct femoral plates; antenna rarely serrate — filiform or short and broad; tarsi rarely lobed; last segment often as long as preceding 3 or 4 together (Dryopoidea); tarsi 4-segmented; fore and mid tibiae broad, spinose externally; antenna short and thick (p. 36). . . . . Heteroceridae
18. Fore coxa projecting, with large exposed trochantin; antenna strongly pectinate or flabellate; hind coxa with well-developed femoral plates (p. 36) . . . . Callirhipidae
    If fore coxa projecting, femoral plate of hind coxa incomplete or lacking 19
19. Metasternum with well-marked transverse suture; abdomen with 2 basal visible sternites fused with suture partly lost; prosternal intercoxal process deeply received into mesosternum; prothorax normally not movable on mesothorax; antenna short, serrate (Buprestoidea) (p. 37) . . . . Buprestidae
    Metasternum lacking transverse suture; suture distinct between abdominal sternites 1 and 2; prosternal intercoxal process not deeply received into mesosternum; prothorax more or less movable on mesothorax; tarsi rarely with more than 1 segment with adhesive lobes below; antenna only sometimes conspicuously serrate. . . . 20
20. Hind coxa almost always with distinct and complete femoral plates; usually 5 visible abdominal sternites — 1st in same plane medially as metasternum; fore coxa rounded; prothorax with acute hind angles; prosternal intercoxal process received movably into a recess in metasternum (Elateroidea). 21

13

Hind coxa with femoral plates very narrow, incomplete or absent; fore coxa large, projecting; usually 6 or 7 abdominal sternites, 1st not in same plane as metasternum; prosternal intercoxal process not or slightly received into metasternum; hind angles of prothorax more or less obtuse (Cantharoidea) 22

21. Antenna inserted close to eye; labrum free; abdomen with 1–3 or 1–4 visible sternites more or less fused (p. 38) . . . . Elateridae
Antenna inserted distant from eye; labrum not visible; abdomen with 5 visible sternites more or less fused (p. 39) . . . . Eucnemidae

22. Without luminous organs; sexes usually similar; hind coxa with femoral plate obsolete 23
With luminous organs in at least 1 sex; sexual dimorphism often strong; hind coxa of ♂ usually with distinct femoral plate (p. 40) . . . . .Lampyridae

23. Trochanters short, obliquely joined to femora; ventral lobe of tarsal segment 4 bilobed; mid coxae close; antennal insertions not very close (p. 40) . . . . .Cantharidae
Trochanters long, transversely joined to femora; lobe of tarsal segment 4 entire; mid coxae well separated; antennal insertions close (p. 41) . . . . Lycidae

24. Distinct spiracles on abdominal segment 8; fore coxa usually projecting; tarsi 5-5-5 segmented; hind coxa often excavate (Bostrychiformia) 25
Spiracle of abdominal segment 8 non-functional; if front coxa projecting, tarsi 5-5-4 or false 4-segmented; hind coxa never truly excavate (Cucujiformia) 27

25. Prothorax nearly always produced forward over head in a hood-like fashion; tarsi with segment 1 very small or trochanters elongate and with an almost transverse junction with femora; ocelli absent (Bostrychoidea) 26
Prothorax never hood-like; tarsi never with segment 1 very small; trochanters normal, their junctions with femora oblique; 1 or 2 ocelli present (Dermestoidea) (p. 41). . . . . Dermestidae

26. Trochanters more or less squarely truncate; antennal insertions close on frons; antenna 11-segmented (p. 42). . . . . Ptinidae
Trochanters obliquely joined to femora; antenna not inserted close on frons; antenna usually with less than 11 segments (p. 42) . . . . Bostrychidae

27. Tarsi variable − if false 4-segmented on all legs, aedeagus with articulated parameres; if head rostrate, tarsi 5-5-4 at least in ♂; if prothorax without distinct margins, tarsi 5-5-4; otherwise 5-5-5 or 4-4-4. 28
Tarsi false 4-segmented on all legs; aedeagus never with articulated parameres; prothorax often without side borders. 44

28. Tarsi always 5-5-5; front coxae projecting (Cleroidea); empodium may be conspicuous 29
If tarsi 5-5-5 front coxa rounded or transverse or aedeagus of 3-lobe type; if fore coxa projecting, tarsi 5-5-4; empodium not conspicuous (Cucujoidea) 30

14

29. Antenna usually clubbed; tarsi never simply 5-segmented and filiform (p. 43). . . . . Cleridae
    Antenna usually filiform, rarely serrate; tarsi simply 5-segmented and filiform (p. 45) . . . . Melyridae
30. Tarsi never 5-5-4 in both sexes; fore coxa never projecting; antenna usually clubbed (Clavicornia = Cucujoidea I). 31
    Tarsi 5-5-4 in both sexes (or 4-4-4); fore coxa usually projecting; antenna usually not clubbed (Heteromera = Cucujoidea II) 37
31. Fore and mid coxae rarely transverse; if tarsi 5-5-5 in both seces, maxilla with 2 lobes and abdomen with 7 pair of spiracles. 32
    Fore and mid coxae very transverse with exposed trochantins; tarsi 5-5-5 or 4-4-4; elytra often truncate (p. 46) . . . . Nitidulidae
32. Antenna 8–10 segmented with 3-segmented club; tarsi 4-4-4; fore coxa transverse, cavity open behind and internally (p. 48). . Cisidae
    Antenna 9–11 segmented with or without 3-segmented club; if head markedly deflexed and prothorax hooded, tarsi rarely 4-4-4 . . . 33
33. Tarsi 4-4-4 (rarely 3-3-3), if 4-4-4 antenna never filiform. 34
    Tarsi 5-5-5 at least in ♀, or if 4-4-4 antenna filiform 35
34. Body round or ovate and evenly convex; antenna short; elytra with sloping margins, rarely margins somewhat flattened (p. 49) . . . . Coccinellidae
    Body widened posteriorly, not evenly rounded and not evenly convex; prothorax often lobed at anterior angles; antenna usually more than ½ as long as body; elytra often with flattened margins (p. 50) . . . . Endomychidae
35. Antenna always clubbed; mid coxal cavity closed externally by episternum; tarsi with segments 2 and 3 lobed below 36
    Antenna usually filiform; mid coxal cavity open; tarsi with segments 2 and 3 not lobed below and 1 shorter than 4; 5-5-4 in male or 4-4-4 in both sexes (p. 47) . . . . Cucujidae
36. Fore coxal cavity open behind; pronotum with more or less distinct prebasal impressions; body narrow; antennal club often asymmetrical (p. 48). . . . . Languriidae
    Fore coxal cavity closed behind; pronotum lacking prebasal impressions; body ovate not extremely narrow; antennal club symmetrical, 3-segmented (p. 48). . . . . Erotylidae
37. Tarsi 5-5-4 in both sexes; mesepimeron reaching coxal cavity; fore coxa projecting above level of prosternum 38
    Tarsi 4-4-4 in both sexes (rarely 3-3-3); mesepimeron not reaching mid coxal cavity; fore coxa not projecting above level of prosternum; antennal insertion usually hidden under side margin of front (p. 50). . . . . Colydiidae
38. Abdomen with segments 1–3 immovable; tarsal claws never appendiculate; fore coxal cavity closed behind. 39
    Abdomen with segments movable; tarsal claws sometimes appendiculate; fore coxal cavity often open behind 41
39. Fore coxa not projecting; prosternal intercoxal process wide; tarsal claws toothed or simple 40

Fore coxa projecting; prosternal intercoxal process narrow; tarsal claws simple (p. 50)......................... Lagriidae

40. Tarsal claws simple; side of frons strongly produced over antennal insertion, projecting into eye; often wingless or with clubbed antenna (p. 51)......................... Tenebrionidae

Tarsal claws pectinate; side margin of frons not strongly produced; usually winged; antenna filiform (rarely pectinate) (p. 54) .... Alleculidae

41. Tibial spurs almost always simple; side border of prothorax usually obsolete; base of prothorax usually narrower than elytra......42

Tibial spurs pubescent; prothorax with borders and as broad at base as elytra; body tapered and acute posteriorly (p. 52)....Mordellidae

42. Antenna filiform to serrate; if tarsal claws serrate, they have long appendages below; prothorax much narrower than elytra 43

Antenna flabellate (at least in ♂); tarsal claws serrate but without long lobes; prothorax at base almost as wide as elytra (p. 53) .... Rhipiphoridae

43. Head feebly deflexed, not sharply constricted to a narrow neck; eye emarginate; elytron with vein-like ribs or 2 penultimate tarsal segments lobed below; mesepisterna not nearly meeting in front of mesosternum (p. 54) .... Oedemeridae

Head strongly deflexed, sharply constricted at neck; eye entire; only penultimate tarsal segment lobed; mesepisterna meeting in front of mesosternum (p. 55) .... Anthicidae

44. Head without a snout; gular sutures separate; antenna not clubbed 45

Head with a snout; gular sutures fused; antenna usually elbowed and clubbed.......................................47

45. Body more or less slender; elytra usually covering abdomen unless greatly abbreviated; antenna usually simple; prothorax often without side margin.............................46

Body very short and stout; elytra truncate and exposing pygidium; antenna serrate; prothorax with distinct side margin (p. 58) .... Bruchidae

46. Antennae often inserted on tubercles, capable of being reflexed backwards over the body, often longer than body; all tibiae with 2 spurs; mesonotum often with stridulatory structure (p. 55) Cerambycidae

Antennae not reflexible back over body, usually less than 2/3 as long as body if on tubercles, tibial spurs 0 or 1; mesonotum rarely with stridulatory structure (p. 59)................Chrysomelidae

47. Maxillary palp rigid; labrum never truly free; gular sutures distinct; prothorax often without side border...................48

Maxillary palp flexible; labrum distinct and separate; gular sutures more or less obsolete; prothorax with side border more or less developed (p. 63) .... Anthribidae

48. Maxillary palp 2–3 segmented; labial palp inserted apically or dorsally; antenna often geniculate, club compact.................49

Maxillary palp 4-segmented, labial palp inserted ventrally on mentum; antenna never geniculate, club 3-segmented (p. 64) ...Attelabidae

16

49. Antenna clubbed, often geniculate, or trochanters elongate; body not extremely slender; snout usually curved downward; sexual dimorphism rarely conspicuous 50
    Antenna straight, not or slightly clubbed; labial palp minute, inserted ventrally in deep pit; body elongate; snout directed straight forward; sexual dimorphism often striking (p. 64) . . . . . . Brenthidae
50. Antenna not geniculate, or trochanters long; ventral surface of mentum with a distinct seta or tuft on side (p. 65). . . . . Apionidae
    Antenna geniculate; trochanters rarely long; venter of mentum lacking projecting setae (p. 66). . . . . Curculionidae

SYSTEMATIC TREATMENT

Following is an outline of the classification used in this handbook. Only groups discussed in this book are listed below.

Suborder Archostemata
Family Cupedidae
Suborder Adephaga
Family Rhysodidae
Family Carabidae
(incl. Cicindelinae)
Family Dytiscidae
Family Gyrinidae
Suborder Polyphaga
Superfamily Hydrophiloidea
Family Hydrophilidae
Superfamily Histeroidea
Family Histeridae
Superfamily Staphylinoidea
Family Silphidae
Family Staphylinidae
Superfamily Scarabaeoidea
Family Lucanidae
Family Passalidae
Family Scarabaeidae (with Trogidae, Geotrupidae)
Superfamily Dascilloidea
Family Helodidae
Superfamily Rhipiceroidea
Family Callirhipidae
Superfamily Buprestoidea
Family Buprestidae
Superfamily Elateroidea
Family Elateridae
Family Eucnemidae
Superfamily Cantharoidea
Family Lampyridae
Family Cantharidae
Family Lycidae

17

Superfamily Dermestoidea

Family Dermestidae

Superfamily Bostrychoidea

Family Bostrychidae

Superfamily Cleroidea

Family Cleridae   Family Melyridae

Superfamily Cucujoidea

Family Nitidulidae	Family Rhizophagidae
Family Cucujidae (with Passandrinae)	Family Languriidae
Family Erotylidae	Family Coccinellidae
Family Endomychidae	Family Colydiidae
Family Lagriidae	Family Tenebrionidae
Family Mordellidae	Family Rhipiphoridae
Family Oedemeridae	Family Anthicidae

Superfamily Chrysomeloidea

Family Cerambycidae	Family Bruchidae
Family Chrysomelidae	

Superfamily Curculionoidea

Family Anthribidae	Family Attelabidae (also Belidae)
Family Brenthidae	Family Apionidae
Family Curculionidae (incl. Scolytinae, Platypodinae)	

Suborder Archostemata
Family CUPEDIDAE

This may be the most primitive living family of beetles, in the sense that it has characters which are quite generalized and simple and which appear to suggest ancestry to other groups of beetles. Quite old fossils in this group have been found. These insects are slender, parallel-sided and flattened above. The antenna is slender, filiform and nearly as long as the body. Pronotum and elytra have ribbed and somewhat reticulated surface. The beetles measure 10—15 mm in length. They are usually blackish or grayish brown. They occur mainly in rotten wood and adults are attracted to light. The larvae are grub-like, without posterior appendages. They bore in rotten wood. The beetles are said to be predaceous in the larval stage. The adults are relatively slow-moving beetles. Most of the species belong to the genus *Cupes* and the number of species is small.

Suborder Adephaga
Family RHYSODIDAE

Members of this family are small, narrow, shiny black insects with short legs, and are 5—8 mm in length. The antenna is short, and the head, pronotum and elytra are usually ridged and furrowed longitudinally to a considerable degree. These beetles are sluggish and little is known of their food or life history, but they are related to purely predaceous beetles. They are usually found under the bark or in the interior of old rotten logs. The larva is grub-like. The genera *Rhysodes* and/or *Rhysodiastes* occur in New Guinea.

CARABIDAE (Carnivorous Ground beetles)

In Papua New Guinea the name "ground beetle" for this family is somewhat misleading as probably a third of the species live in foliage, on tree trunks, among epiphytes and mosses, or under bark. The subfamilies Carabinae and Harpalinae contain 700 described New Guinea species and certainly many more await discovery and description. Darlington (1952—1971) has studied this group intensively and his work should be consulted, as it contains keys to the identification of all the known species. He has classified the group into mesophiles (a third of the species), living on ground, hydrophiles (a third of the species), existing in the vicinity of water, and arboreal forms (a third), living in standing vegetation.

The beetles vary somewhat in shape, some being oval, while species of numerous genera are parallel-sided. They tend to be rather flattened and have forward directed prominent mandibles. The legs are long, they are fast moving, predaceous insects which hunt and feed on other insects, worms and larvae. They presumably play an important role in the control of insect populations. In many species the wings are well developed and

19

they may take flight readily; some come to light. Darlington has shown that mountain species have a tendency for reduction in wing size and many in the higher regions are wingless. While most of the beetles are uniformly dull or shiny black in color, some have yellow or buff patterns on the elytra. Other species, particularly those having arboreal habits have shining green, bronze and coppery colors. The larvae are also active and predaceous.

In Papua New Guinea the large northern Carabini are lacking and the familiar black ground beetles, Pterostichini, which are commonly found under stones or logs in Australia and the temperate countries of Europe or America, are in the minority, though several genera are represented. The New Guinea carabids tend to be small and rarely exceed 25 mm in length. They are not usually encountered under stones or logs lying on the ground, but large numbers of species inhabit the leaf litter on the forest floor. Some genera, as *Tachys* and *Bembidion* have many very small species living in the shingle, gravel beds beside rivers and streams, often running over the mud beside ponds and swamps.

The tribe Pentagonicini is another group of ground beetles which contains several genera which live on trees. Members of the genus *Scopodes* are, however, denizens of the ground and also are encountered on fallen logs. They are very small, usually being under 5 or 6 mm in length. They have a characteristic shape with a prominent prothorax, oval elytra, with particularly large eyes. They are diurnal insects which may sometimes be seen swarming on moist roadside cuttings, particularly if these are moss-covered, and at other times running over wet fallen timber. The genus is also known to occur in Australia and New Zealand. A characteristically shaped group of ground beetles are the Odacanthini, which can be recognised by the elongated prothorax which is both long and narrow, and narrower indeed than the width of the head. These insects appear to frequent wet, marshy localities but are sometimes found around lights.

The Scaritinae contain parallel-sided beetles about 2–8 mm in length with fore tibiae which are thickened and flattened. This is a specialization for digging. The beetles live in holes which they dig in soft ground and are numerous in the Fly and Sepik River valleys as well as elsewhere in the lowlands. They appear to be nocturnal and often come to light. Most of them belong to the genus *Clivina*.

The largest group in Papua New Guinea is the Agonini with at least 160 species belonging to 21 genera. Many of these genera are found only in New Guinea. Some of the arboreal forms are quite brightly colored. *Notagonum* (mostly lowland, I, II) and *Altagonum* (montane, III, IV) include many species. *Colpodes* is shown in Fig. 3.

---

*Asterisked genera are apparently endemic to the island of New Guinea or at least to the Papuan Subregion. See page 2 for definition of altitudinal zones, I-IV.

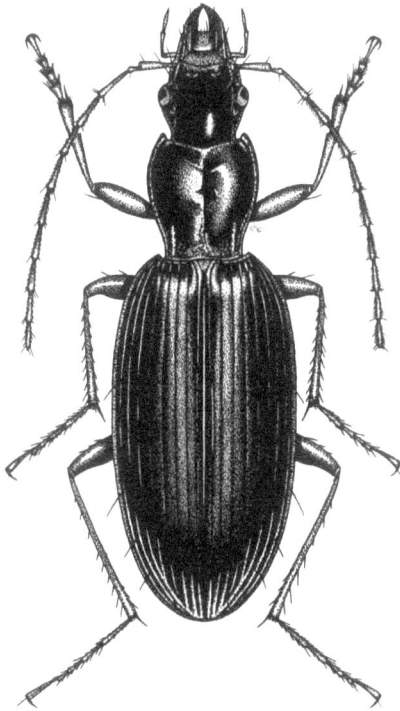

Figure 3 *Colpodes habilis* Sloane (Carabidae: Agonini).

The tribe Lebiini includes a large number of species which have a generally rectangular outline and are somewhat flattened. The elytra of many are obliquely truncated. Many are arboreal, living on the foliage or on the bark of trees. The genus *Catascopus* includes a number of species which live on and under the bark of dead trees. Most of them are handsome, shining, blue-green or purple in color and quite large in size, ranging from 10 to 20 mm. They have prominent eyes and long legs and are very active insects, quickly running for cover when a piece of bark is pulled from a log, and flying readily if cornered. *C. elegans* Weber is 10 mm long and is a metallic green color. It is very common in lowland rain forest, whilst *C. wallacei* Saunders has rather purplish colored elytra with a bright coppery red prothorax. It is 18 mm long and seems to be quite common in the highlands. Another important genus is *Demetrida* which contains some 160 or so species. These are frequently encountered on the foliage of rain forest trees. Their colors vary from a green, with various shining reflections, to brown and black. They are not particularly large insects, being some 5–8 mm in length.

CICINDELINAE – this subfamily, the tiger beetles, is usually called a separate family, Cicindelidae, in older works. They are mostly 8–12 mm long and the elytra are usually dull black patterned with an intricate design of white and yellow lines or spots. The head carries a formidable and conspicuous pair of downwardly directed mandibles which can deliver a sharp nip if the insects are carelessly handled. Some species are commonly seen on the exposed clay around villages or in similar situations along tracks or stream beds. They fly very readily and rapidly and usually move one or two meters ahead of an intruder then circling to land at his back. At times they occur in large swarms in favorable locations. Many species hunt in the bright daylight and can be observed tracking down and catching in their jaws small flies and other insects. Some species are nocturnal and of these many come to light. They may also be observed hunting the insects which are attracted to light where the shadows begin to deepen. The largest group of species belong to the typically terrestrial genus *Cicindela*.

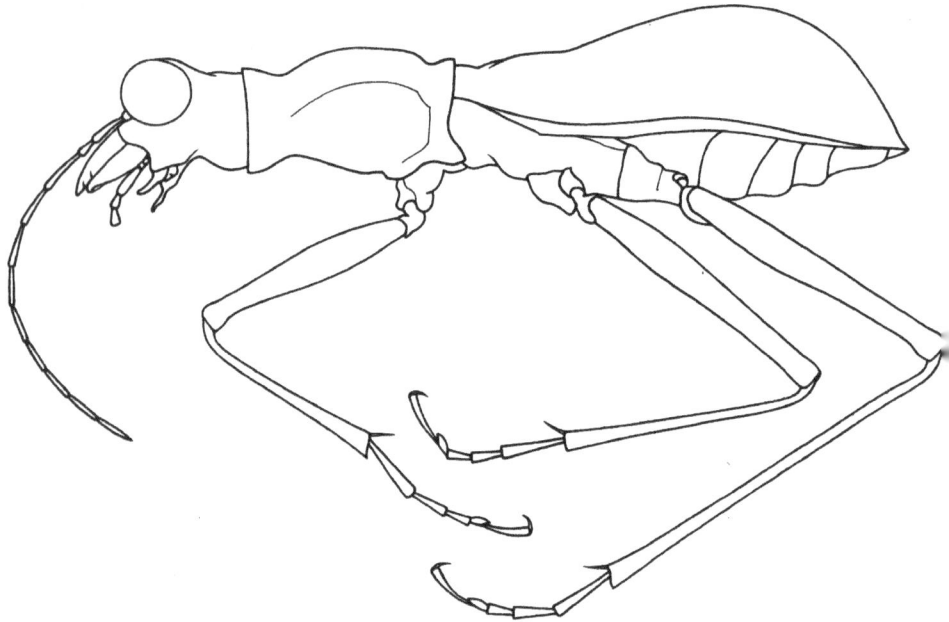

Figure 4 *Tricondyla aptera* **Olivier (Carabidae: Cicindelinae).**

Tricondyla aptera Olivier is a large black ant-like beetle, wingless, but with long legs. It is often seen roaming over rubbish where new gardens are being worked, climbing trees and running over the fallen branches. Besides being abundant in the lowlands of New Guinea, it is also found in Indonesia and North Queensland.

Therates labiatus Fabricius (Pl. 1a) and *T. basalis* Dejean are two shining black tiger beetles with large yellow jaws which walk on the leaves of broad leaved plants such as gingers in clearings in rain forest or beside paths, more at lower altitudes. They have a heavy and slow flight when disturbed.

Distipsidera contains two species which may be found hunting over the trunks of standing trees in the savanna regions of the trans-Fly, and probably elsewhere in southern New Guinea. These are large insects with conspicuous blobs of yellow coloration on their elytra and bodies. *Megacephala* species also occur only in southern savanna country.

The larvae of tiger beetles construct tunnels in clay banks and similar surfaces and can sometimes be drawn out by pushing a small thread of grass into the entrances. This they seize with their large mandibles which are normally employed in the capture of their prey, the head of the larva being held in the entrance to the burrow awaiting the passage of an ant or other small insect which they quickly seize. These larvae have certain anatomical specialisations which include various bristles and a humped back which effectively secure them to the walls of their burrows, preventing predators from pulling them out, and also providing them with greater purchase and strength when pulling prey into the burrow. In Europe these larvae are called "penny doctors" by children.

Family DYTISCIDAE (Carnivorous water beetles)

These beetles are oval or boat-shaped, usually black in color, often smooth and convex and streamlined. They vary in size from a few mm to over 30 mm in length. The adults and their free-living larvae are to be found particularly in still waters of ponds and swamps, but may also exist in flowing water. There are many species in Papua New Guinea and in the Sepik area some of the larger species are caught in small fish nets and are valued as food.

The adults swim rapidly using the specialised, paddle-like hind legs for propulsion. They have to come to the surface frequently for air which is taken into the body through terminal spiracles on the abdomen, and a bubble of air is also stored between the elytra and the abdomen. The beetles have well-developed wings and fly at dusk and during the night, and by this means travel to new water supplies such as recent rain ponds. In areas of the country where there is much standing water, large numbers may sometimes be attracted to light. The larvae are long and slender, with long mandibles and terminal gills.

Genera known from New Guinea include *Cybister* (large and very broad), *Rhantus, Copelatus, Canthydrus, Hyphydrus, Hydaticus* (Pl. 5b). (also Pl. 5c).

Family GYRINIDAE (Whirligig beetles)

This is another family of water beetles of similar shape to the Dytiscidae, but in these insects the fore legs are particularly long and extend in front of the body. The eyes are highly specialized so that the insect may receive images from both above and below the water line on which it floats. Commonly the fore femur is long and has a sharp angular bend to the tibia which protrudes in front of the insect for some distance. The mid and hind legs are shorter and more paddle-like. The beetles swim on the surface of small streams and occasionally standing pools. They tend to be gregarious and groups of 10 or more may occupy the same corner of the pool. They favor shady, still or running water where they ripple the surface with their constant movement, tracing out irregular circles and zigzag patterns. When alarmed they immediately dive and shelter under stones on the bottom of the pool. Some of the insects are quite large, being 20 mm in length, and are often quite broad. They are uniformly olive green, brown or black in color. Larvae are slender with lateral gills. Genera in New Guinea include *Macrogyrus, Dineutes (& D. Merodineutes* Pl. 5a), *Paragyrinus,* and *Gyrinus.*

Superfamily Hydrophiloidea
Family HYDROPHILIDAE (Scavenging water beetles)

This is another family consisting largely of water beetles, which, unlike the Dytiscidae and Gyrinidae, are primarily scavengers and feed on vegetable matter including algae. The beetles are usually black or brown and are convex-oval in shape. Many superficially resemble Dytiscidae but are usually more convex. They vary in size from 2 or 3 to 30 or 40 mm in length. The maxillary palp, an appendage of the maxilla behind the mandible, is usually long and may project in front of the head further than the relatively short antenna. This is a good means of differentiating these insects from the other water beetles. The larger species have to break the water surface to acquire air stores for respiration, although many of the smaller species may gain enough oxygen from the bubbles on the surface of plants and weeds in the ponds where they dwell. All aquatic species have hairs on some body surfaces which retain air for breathing. The beetles fly freely at night, being attracted to light, and some may even be found on the wing during the day.

Some species, such as those of *Sphaeridium, Dactylosternum* and *Cercyon* live in dung and do not occur in water. Aquatic New Guinea species belong to *Hydrophilus* (Pl. 5d), *Coelostoma, Anacaena, Helochares, Hydrochus, Cryptopleurum, Enochrus, Laccobius* and *Sternolophus.* Species of *Hydrophilus* are much larger than the others. Most of the aquatic species produce a sort of cocoon to house their eggs.

Superfamily Histeroidea
Family HISTERIDAE

This family includes a number of small, shiny, smooth black or rarely green beetles. The size varies from 2–10 mm. The elytra and body are hardened, and as the legs retract into grooves beneath, the disturbed beetles may appear to resemble small hard black seeds. They are usually flattened and squarish or oblong in shape, and most of the species look rather similar. They are mostly found under bark, but a large species congregates in the substance of decaying sago palms, and others are encountered in carrion, dung, fungi or stored fooods. Larvae and adults are predaceous and feed on other small insects and larvae. Those living under bark are flatter than others. Some occur in ant or termite nests and may have a symbiotic relationship with their hosts. New Guinea genera include *Plaesius* (Pl. 5 e), *Eblisia, Hololepta, Teretriosoma, Epiechinus* and **Nicotikis.*

Superfamily Staphylinoidea
Family SILPHIDAE (Carrion beetles)

This family is not very conspicuously represented in New Guinea. The beetles are moderately large, more or less flattened, oblong, with leathery elytra which sometimes incompletely cover the abdomen and the hind wings. In some (*Ptomaphila*) the elytra are raised into a number of tubercles which resemble small blisters, while raised longitudinal ridges are also often developed.

Diamesus osculans Vigors (Pl. 5f) is widely distributed in the Indo-Australian region, including New Guinea mainland and New Britain. Its general color is black, but the elytra have two large yellow marks. It is quite a large insect, sometimes reaching 25 mm in length. Like other members of the family it inhabits carrion. It is occasionally attracted to light, making a loud noise in flight and giving off an unpleasant odor. Species of *Nicrophorus* are deeper bodied and usually black with red markings.

Family STAPHYLINIDAE (Rove beetles)

The beetles belonging to this family have the elytra much shortened so that two-thirds to three-fourths or more of the abdomen is exposed. This gives the insect an ant-like appearance which is enhanced by the fact that many are very active creatures which may run over the ground with the speed and restlessness of ants. The flexible abdomen is sometimes extended vertically and moved vigorously around in a threat display when the beetles are alarmed. The hind wings are folded in a complex and compressed fashion beneath the abbreviated elytra. In spite of this they may be very rapidly uncovered and the beetles can take to flight with great rapidity. In bright sunshine some species may be seen flying and settling on leaves and around fungi in forest clearings. Such insects often run over tree trunks and foliage with all the mannerisms of a small wasp or ant. Many of the New Guinea species have the terminal segments of the antenna white or yellow and their actions and behavior and waving of the antennae may give the impression of a small wasp. The beetles vary in size from less than 1 mm to 25 mm. The majority are, however, small. The larvae are often found in association with the adults and they have a general superficial resemblance to the latter.

New Guinea possesses a very rich fauna of these insects and undoubtedly many species await scientific description. The adults and larvae are predators feeding on mites and other small creatures. They often swarm in damp decaying vegetable matter, such as piles of freshly pulled garden weeds, and a handful of such material may produce hundreds of specimens of several species. They are also frequently found in rotting fruit — pawpaws or bananas and figs. Other species congregate in fungi, both in fresh and decaying stages. Some species are attracted by dung and carrion, while many are found on the ground and among the shingle of river beds.

The Xantholinini are flattened rove beetles with large forwardly directed mandibles, which are found under the bark of trees and in rotting wood. A genus which inhabits such situations is *Priochirus* (Pl. 5g). This group includes common and conspicuous insects often found in groups under damp bark of dead trees. They are shiny black with a heavily indented and grooved head and prothorax which gives them a massive appearance. Some are fairly large, 12 mm or more in length. They are sluggish compared with many rove beetles, and presumanly prey on other insect larvae living in the same situations. *Dinoxantholinus* is another genus in this group.

Paederus is a genus of ant-like beetles about 10 mm in length which live in association with running water and are often visible on boulders in the spray zone. These insects occasionally discharge an irritant fluid when touched, and cases have been recorded of skin rashes resulting from contact with the beetles.

Creophilus albertisi Fauvel (Pl. 5h) is a large staphylinid with a bright orange-red head and metallic blue and green wing covers, approximately 16—20 mm in length, which occasionally comes to light, and which can discharge a peculiar pungent scent when frightened.

The genera *Philonthus* and *Hesperus* (Fig. 5) contain a large number of species of typical staphylinid appearance, and are particularly attracted to rotting fruit and sometimes to the fermenting sap which exudes from wounds on tree trunks. Many are brilliantly colored in shades of iridescent purple, bronze, red, green and blue. They are very active insects which rapidly take to the wing, and they also run with great speed to shelter when disturbed, and are often difficult to catch for this reason.

Stenus is a genus of relatively long-legged beetles of slight, slender proportions but with large bulging eyes and a prominent forward projection of the mandibles and labrum between the eyes. They have dense punctures on the body and elytra. Many are of shining metallic colors. They live in marshy places and are sometimes beaten from foliage near water. One small black species,

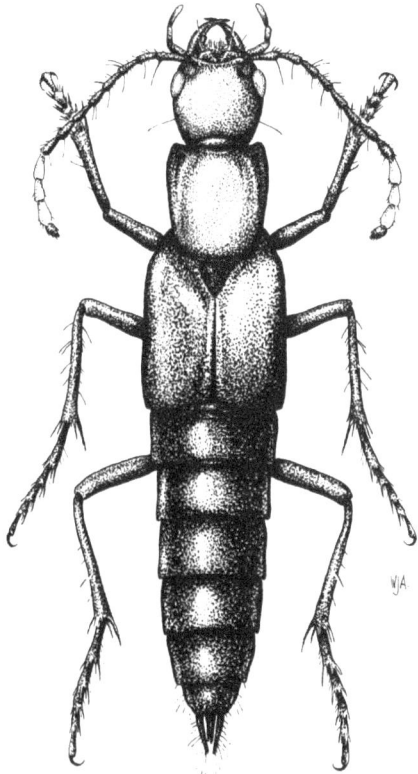

Figure 5 *Hesperus abnormis* Cameron (Staphylinidae).

S. sepikensis Last, occurs in vast numbers on the muddy banks of drying up pools in the Sepik and Fly River valleys. Here, it runs over the mud and resembles an ant.

Among other genera occurring in New Guinea are *Leucitus, *Scelotrichus, *Stichostigma, Lispinus, Osorius, *Pseudocraspedomerus, Aleochara, Thinocharis, Oxytelus, *Troposandria, Leptochirus, Eleusis, Thyreocephalus, Edaphus, Erchomus* and **Pachypelmopus*.

The related family PSELAPHIDAE contains small beetles of much shorter body, also with short elytra, and mostly living in ant nests.

Superfamily Scarabaeoidea
Family LUCANIDAE (Stag beetles)

This family includes some of the larger and more handsome beetles of New Guinea. There are not a large number of species here, and the fauna is not as rich as it is in Southeast Asia. The main recognition character is the very large mandible of the males of most species. Often the mandible is about one-half as long as the body, but may be more or less. The head of the male is about as broad as the prothorax, and the body is more or less flattened. The females are smaller than the males and have much smaller mandibles. The jaws of the males vary in size within the species, more or less correlated with size of body (and probably also with the success of the larva in obtaining adequate nourishment). Males may be called 'major' when the mandibles are large and 'minor' when jaws and body are relatively small. Some males may be one-half as large as others of the same species. The antenna is geniculate, with a flattened club. Most stag beetles are black, brown or bronzy, but some are strikingly marked with dark brown and yellow, or otherwise. Most kinds are smooth, but some are striate or pubescent. Adults are mostly from 15 to 50 mm long, but some are smaller and some still larger.

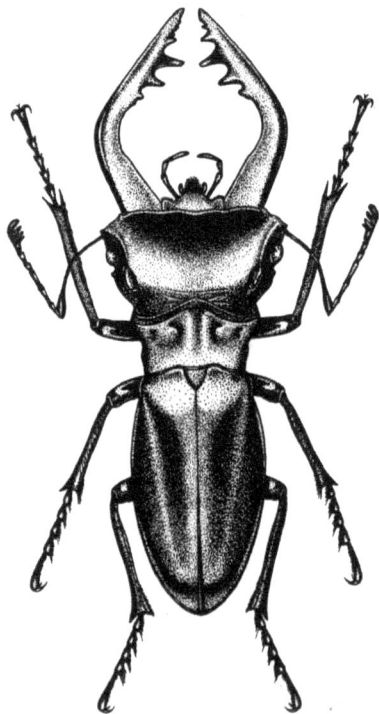

Figure 6 *Cyclommatus sumptuosus* Moellenkamp **(Lucanidae).**

The larvae are whitish, long and arched, more slender and softer-bodied than most scarabaeid larvae. Usually the larvae are in rotting stumps and logs, feeding on decaying wood. The adults are found on trunks of trees, where they may lap up juices coming from borings such as by long-horned beetle larvae in the wood of a live tree. They also occur on stumps and logs. Common genera in New Guinea are *Cyclommatus* (Fig. 6) — metallic with very broad head (*C. finschi* and *C. imperator* may have mandible as long as rest of body); *Serrognathus* — black with heavy mandible with sometimes only one tooth; *Prosopocoelus* (or *Metopodontus*) — brown and yellow striped, very smooth, with long slender mandibles; *Neolamprima* (Pl. 1b) — bronzy with vertically upward arched jaws; *Pachystaegus* — medium size, broad body; *Aegus* — relatively short and flat with usually a single tooth on inner side of jaw, and with ribbed elytra (Fig. 7); *Figulus* — smaller, shiny and also

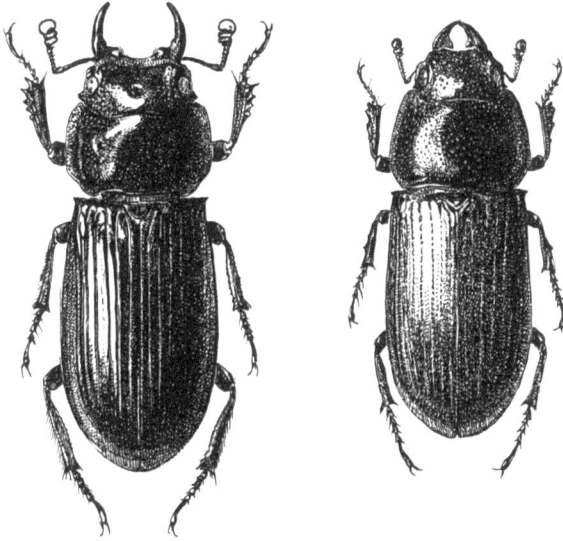

Figure 7 *Aegus gressitti* de Lisle (Lucanidae), male
left, female right.

ribbed, with quite narrow body; and *Gnaphaloryx* — small and short with
smaller mandible.

Family PASSALIDAE

This family is closely related to the Lucanidae. Members are usually black,
but are red-brown when teneral, in cells under old bark of logs. The body
is flat above, the pronotum smooth, often with a fine median groove, and
the elytra are ribbed longitudinally. The antenna is fairly short, with an
open flattened club. The body often has reddish hairs beneath. Size of
body varies from about 12 to 45 mm or more in length.

The larvae are elongate, slightly arched posteriorly, fairly soft-bodied
and whitish, with the gut often showing through the semi-transparent skin.
The larvae feed in old rotting logs and stumps. Often larvae and adults are
seen more or less together. They appear to be somewhat social in habit,
and the adults apparently protect the larvae from predators to some extent.
This is an unusual habit for beetles.

New Guinea genera include *Analaches, Aulacocyclus, Cetejus, Episphe-
noides, Gonatas, Kaupiolus, Pelopides* (Fig. 8), *Leptaulacides, Leptaulax,
Mastochilus, Omegarius, Protomocoelus,* *Pseudepisphenus* and *Tristor-
thus.*

29

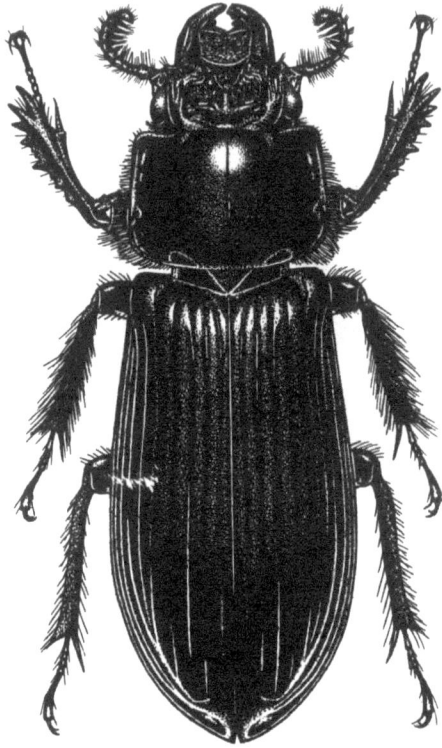

Figure 8 *Pelopides schraderi* Kuwert
(Passalidae).

Family SCARABAEIDAE (scarab beetles)

This family contains a large assemblage of generally stout beetles, some of which are among the largest insects in the whole order. They can usually be recognised by the fact that the terminal segments of the antenna are specialised to form a club of 3–7 moveable plates or lamellae. Further-more, the terminal one or two segments of the abdomen are usually visible and protrude beyond the end of the elytra. The legs are often specialised for digging, with widening of some segments, and expansion and extension of the claws which may carry an elaborate group of bristles and hairs. The hind wings are freely used in flight but are not visible in the resting insect, being folded beneath the elytra. Most species fly at dusk or during the night, but members of the subfamily Cetoniinae are diurnal. The larvae are white grubs, usually arched or nearly forming a circle when in the feeding or resting position. They have a reddish brown head equipped with large jaws. The larval thorax bears three pairs of small legs which, in spite of their size, permit the grub to crawl and dig and burrow speedily into the soil. The larvae live in soil and soft rotting timber or dung. Some species may constitute a serious nuisance in gardens, lawns or pastures, where they feed on the roots. Larvae of many species are palatable and sometimes eaten by man.

APHODIINAE – This subfamily contains small parallel-sided or cylindrical black or gray beetles, 2–6 mm long. The elytra are often smooth, but in some have longitudinal ridges and grooves. These beetles live in dung and rotting vegetation and fly at night, when large numbers may be attracted to light.

COPRINAE or dung beetles have an oval or round to subsquarish shape. They are usually uniformly black in color but some have dark greenish reflections and others have grayish scales which produce a somewhat mottled pattern. In Papua New Guinea they are usually under 12 mm in length and some are quite small. The legs carry conspicuous claws and bristles and the front of the head is often specialised, particularly in the males, to form

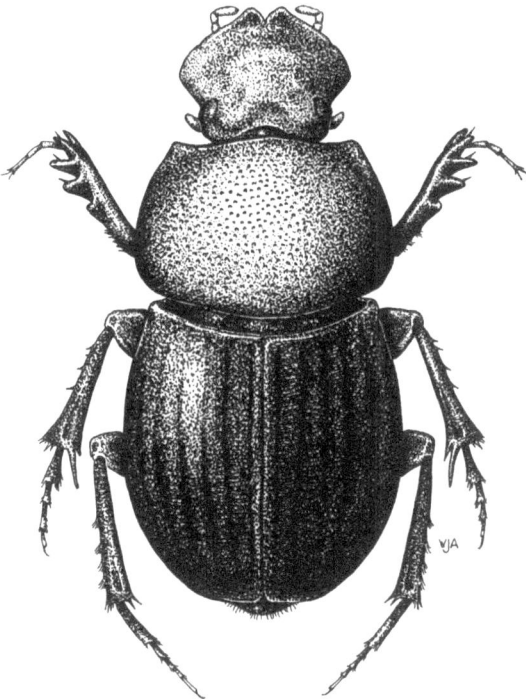

Figure 9 *Onthophagus latinasutus* Arrow (Scarabaeidae: Coprinae).

more or less prominent horns. The front of the head is also expanded forward to form a flange or rim overhanging the mouthparts. These modifications are specialisations so that the head may be used as a wedge or tool in the formation of pellets of dung, which form the food for both adults and larvae. The adults are attracted to the excrement of mammals and man, and may be found on the underside of a deposit of excrement. They may construct tunnels deep into the underlying clay or soil and carry the pellets down to a nest where eggs are laid and the larvae later develop. These dung beetles, including also the Aphodiinae, play a useful role in removing excreta and burying it. This both reduces fly populations and improves soil. In New Guinea many species also feed on rotten bananas and other decaying fruits, or nuts, lying on the ground in the forest. (See Fig. 9).

MELOLONTHINAE, or chafers, represent another subfamily of the scarabs. They are usually of a rather dull brown or reddish color and are commonly seen flying at dusk or coming to light at night. The beetles lift the elytra vertically above the back of the abdomen when taking off in flight and their flight is characteristically heavy and ponderous, accompanied by a familiar droning sound. In many areas of the highlands, particularly in the Chimbu, Asaro and Kainantu regions, large numbers of a common brown species some 25 mm in length fly at dusk at the end of the dry season after showers of rain. They are caught by using a branch as a flail and collected in bamboos or other containers, being subsequently roasted and relished as an item of food. These flights of large numbers of beetles are accompanied by mating, and subsequently the females lay eggs which are deposited in the pasture. The larvae and adults probably form an important item of diet for pigs, and are an explanation for the rooting of pigs in lawns and town park-land. Species of *Dermolepida* (III) are destructive to lawns, pastures and grass airstrips (Pl. 6b). Other genera are *Maechidius, Heteronyx, Apogonia.*

Several larger species with blotchy white or grayish markings on buff or brown elytra are found commonly in the lowlands. One species shelters by day in the hanging clusters of dead banana leaves, and in some areas of the Western Province of Papua is collected for food.

RUTELINAE is another scarab subfamily more oval in outline than the melolonthines, and with smooth, shiny, often metallic elytra. The hind legs have thickened tibiae and the tarsi have long claws. A common shiny dark green insect is often found clinging to leaves of shrubs and low trees in lowland areas and flies to light at night. In Australia beetles of this subfamily are popularly described as Christmas beetles, and in some north temperate areas members of this group together with the chafers may be called May beetles or June beetles. New Guinea genera include *Lepidiota, Lepidoderma* and *Anomala.*

CETONIINAE form another group of scarabs, moderately large, day-flying beetles, which are attracted to flowering trees and are hence popu-

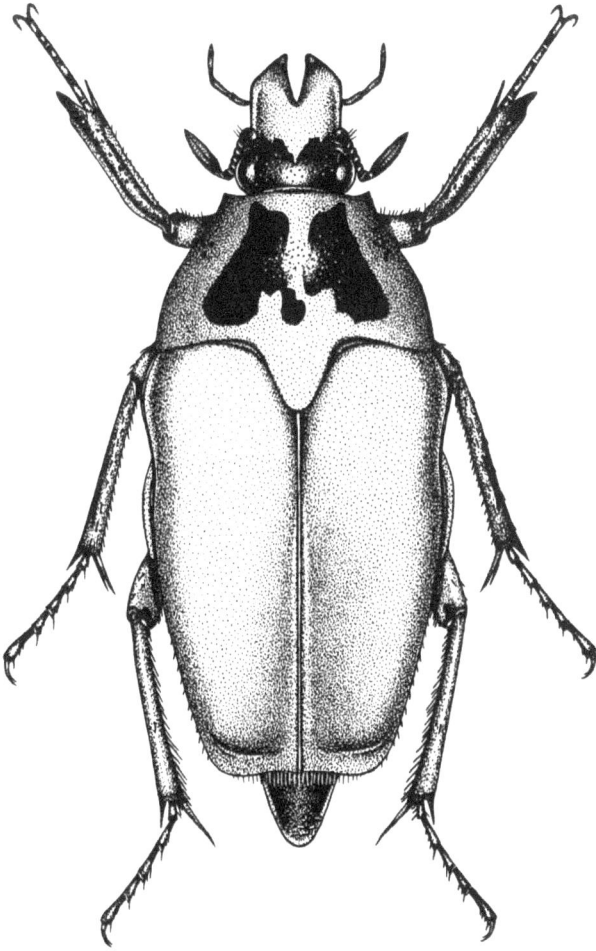

Figure 10 *Lomaptera lutea* (Linnaeus) (Scarabaeidae: Cetoniinae).

larly called flower or rose chafers. They feed on nectar. Their flight is fast and straight or with varying inclination so that they may be difficult to capture. They settle on flowers of forest trees in large numbers but take off rapidly at the slightest sign of an intruder. They are thus much more active on the wing than members of other subfamilies of scarabs. These beetles tend to be rather flattened compared with the more convex and rounded or cylindrical shape of most scarabs. The edge of the elytron is incurved at the level of origin of the hind leg, producing a concavity which is usually clearly visible. The beetle raises the elytra only slightly, and the hind wings function through these lateral emarginations. The antenna has a prominent lamellate club, but the termination of the abdomen is usually covered by the elytra. Many species are brilliantly colored, being shiny with metallic reflections. Green is the most common color, but some are bronzy red or other shades. Members of the genus *Lomaptera* are common beetles and there are many species. In addition to metallic colors some are fawn, black or dark brown and a common species has a conspicuous inverted W-shaped mark on the prothorax (Fig. 10). The elytra, or entire beetles, of metallic members of this group are regularly used in head-bands or other decorations for singsings. *L. jamesi*, with a splash of red across the green elytra, is prominently used for such purposes throughout the Highlands, although the beetle occurs wild only in the southern Papuan area and must have been traded into the Highlands since long ago as part of the traditional trade. Other genera are *Ischiopsopha, Macronota, Glycyphana* and *Protaetia (P. advena* on *Erythrina*).

DYNASTINAE include the rhinoceros beetles. These scarabs are moderate to very large insects which have great importance through their role as pests of agricultural crops. They are stout rounded beetles. The males often have peculiar modifications of the prothorax and head to give them a horned appearance. The horns may show considerable variation in size between individuals of the same species, and in small individuals may be markedly and disproportionately reduced. The function of these horns is poorly understood. It probably relates to fighting among males and may have some function in courtship of the females. *Eupatorus beccarii* Gestro (II) is a large black and reddish species (Pl. 1c).

Xylotrupes gideon L. (II) is a very common and familiar species with a single prominent forward-directed horn which divides near its termination, extending from the apex of the prothorax, almost meeting an upward-directed horn which extends vertically from the head. The female has neither prominence and the prothorax and head are simply rounded in a conventional fashion. The beetle is attracted to light in almost all areas of Papua New Guinea, and at times these beetles may also congregate in large numbers on cassia and poinciana trees. The male particularly is capable of stridulating – making a quite discernable noise by flexing and extending the prothorax against the elytra. This beetle is also a pest in its larval stage. Of even greater importance are *Scapanes australis grossepunctatus* Stemb. (I), which occurs in New Britain and Bougainville, and *S. a. australls* (Boisd) (I), which occurs on the New Guinea mainland. These beetles have two forward projecting horns arising from the prothorax and jutting horizontally in the long axis of the beetle. A central hook is directed upward from the top of the head. A somewhat similar beetle is *Oryctes rhinoceros* L. (I), which is an Asian insect quite recently introduced to New Britain and Manus. These beetles are all serious pests of coconut plantations, the adult insect burrowing into the central growing area of the palm and eventually

34

feeding on the soft pithy material, when it may kill the palm. The only effective remedy is to manually remove the beetle and destroy it, although some natural enemies occur, such as large scoliid wasps. *O. centaurus* Sternb. (I) normally attacks sago palm. Members of the genus *Oryctoderus*, more cylindrical and with short horns, are said to be predaceous in the larval stage on other dynastid larvae. There are many smaller Dynastinae which are also pests, both in the adult and larval forms, burrowing and eating roots, often the roots of food plants such as taro. Most of these beetles are 10–20 mm in length, with a prothorax and head bearing smaller, black upward and forward directed horns, rarely as conspicuous as with the larger rhinoceros beetles. These species are also attracted to light and sometimes large numbers may congregate around street and house lights on wet and warm evenings. *Papuana* species, *P. huebneri*, Fairm., *P. japenensis* Arrow (I) (Pl. 6c) and *P. woodlarkiana* (Montr.) (II) are especially important, eating taro in coastal areas.

HYBOSORINAE includes relatively flattish forms, usually plain brown: *Liparochrus, Phaeochrous* (Pl. 5i). The TROGIDAE include rough, oval beetles 5–10 mm long which usually feed in dead animals, and the GEOTRUPIDAE are more rounded and convex, with longitudinal grooves on elytra, and are dung beetles.

Superfamily Dascilloidea
Family HELODIDAE

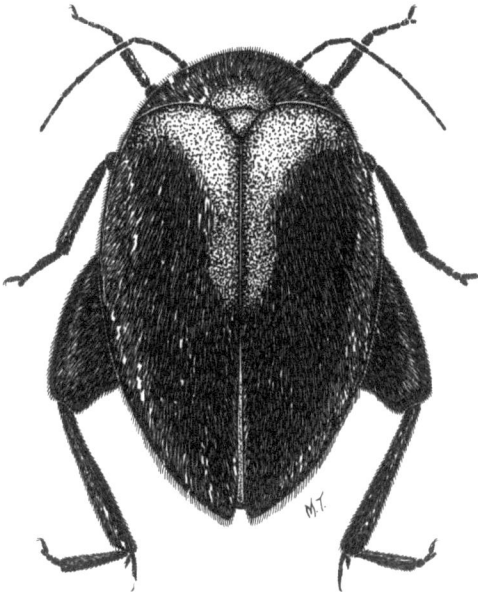

Figure 11 Family Helodidae.

35

Members of this group resemble lady-bird beetles in shape, being almost round in outline in dorsal view, and rather evenly convex above. The color may be pale testaceous to black, and is usually uniform, rather than having spots as in many lady-birds. Size is usually 4–8 mm in length. The larvae are said to live in stagnant water, and to have gills and segmented antenna. The adults often occur in great numbers on felled trees and logs in damp places. An example is illustrated in Fig. 11.

The related family HETEROCERIDAE has flatter and narrower species which are quite small and often mottled brown and paler.

Superfamily Rhipiceroidea
Family CALLIRHIPIDAE

This group of beetles, close to the family Rhipiceridae, is not a large one, but usually one or two species are present in lowland or lower montane areas of New Guinea. These two families, together with the Elateridae and

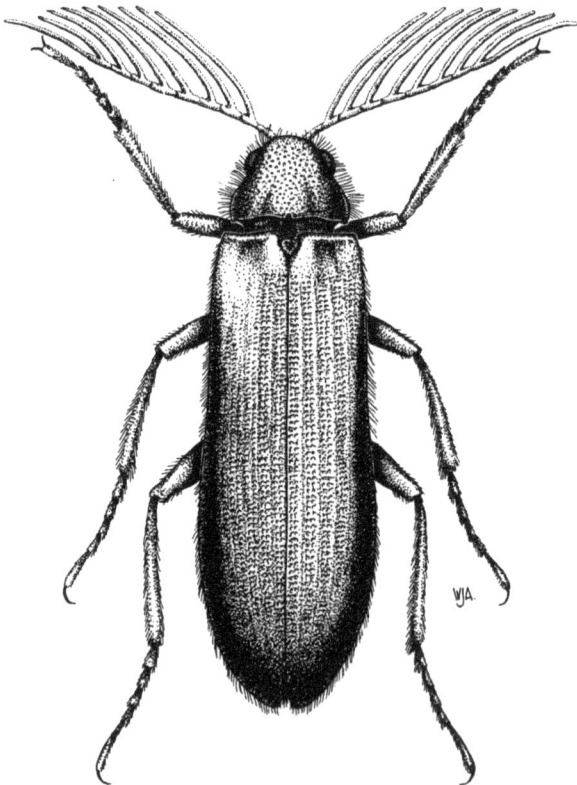

Figure 12 *Callirhipis* sp. (Callirhipidae).

Buprestidae, have been grouped together in the past as "Serricornia" from their frequently serrate or pectinate antennae and similar form, though now divided among several superfamilies. Callirhipids are moderate-sized beetles, mostly 8–15 mm long. Common ones are parallel-sided, moderately convex, rounded behind, with antenna pectinate and legs of modest length. The color is usually reddish brown to blackish pitchy and the surface may be pubescent and mat or glabrous and slightly shiny. The larvae are long, cylindrical and pigmented, living in rotten wood. The genus *Callirrhipis* (Fig. 12) occurs in New Guinea.

Superfamily Buprestoidea
Family BUPRESTIDAE

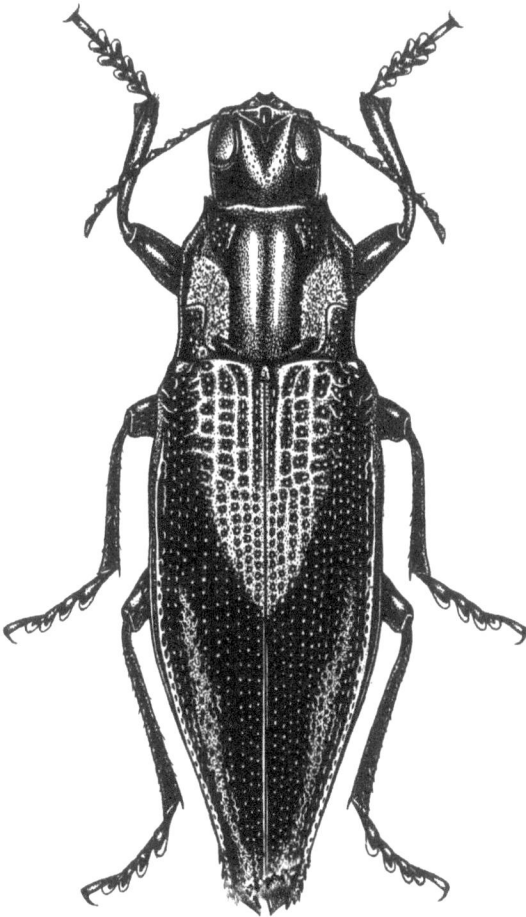

Figure 13 *Cyphagastra timorensis albertisi* Gestro (Buprestidae).

Members of this group are known as metallic wood-borers, jewel beetles or flat-headed wood-borers. Some of the species are very beautiful, having markings of metallic blue, green, gold or coppery, or being entirely metallic above, or over the entire body. The body is usually long, somewhat flattened, and narrowed posteriorly. Some species are very small. Size varies from about 3–30 mm or more. Some of the very small ones are wide and short, scutiform, and are leaf-miners in the larval stage, whereas most species are wood-borers. Large species may bore under the bark of large dying tree-branches, or trunks, and some of the more slender species bore in twigs or stems of herbs. The larvae are strongly flattened, pale, with expanded thoracic region, giving rise to the name 'flat-headed wood borers.' The adults are strong fliers, but they do not usually fly at night as do elaterids. Some workers have considered the elaterids and buprestids to be very closely related, whereas others have place them in different, but adjacent, superfamilies.

Among New Guinea genera are *Agrilus, Anthaxia, Aphanisticus, Belionota, Callopistus, Chalcotaenia, Cisseis, Chrysobothris, Coraebus, Cyphogastra* (Fig. 13; Pl. 1d), **Demostis, Diceropygus, Endelis, *Helferella, Iridotaenia, Melobasis, Merimna, *Metataenia, Metallesthes, Metaxymorpha, Paracupta, Philanthaxia, Polycesta, Stigmodera.*

Superfamily Elateroidea
Family ELATERIDAE (Click-beetles)

This familiar group comprises the click-beetles or snap-beetles, recognized by the muscular articulation between the prothorax and mesothorax by which the beetles snap their bodies to escape or right themselves. The legs are relatively small and weak. The antenna is usually serrate or flabellate. Body length varies from about 3–30 mm or more. Colors are varied, but often mottled brown or black and gray, while some are banded with orange or red, or have large spots. Species of *Alaus* and relatives are very large, somewhat ribbed longitudinally on the elytra, and commonly fly to lights at night in the highlands.

The larvae are long, often quite slender and cylindrical, or somewhat flattened. They may be dark brown, reddish or pale. The slender cylindrical ground-living larvae ('wire worms') may be reddish brown, feeding on roots but free in the ground. Some of the flatter larvae may be pale, and live in the burrows made by other insects in logs and stumps, feeding upon larvae of other beetles or other insects. Thus, some species may be pests of agriculture and others are beneficial from man's standpoint, playing different roles in local food-chains.

Among New Guinea genera are *Agonischius, Alaus* (Fig. 14; Pl. 6a), *Anchastus, Anthracalaus, *Cardiodontulus, Cardiophorus, Compsolacon, Conoderes, Dioxypterus, Elater, Hapatesis, Lacon, Megapenthes, Melanotus, Melanoxanthus, Neodiploconus, *Oxystethus, Propsephus, *Symphostethus, Tetrigus.*

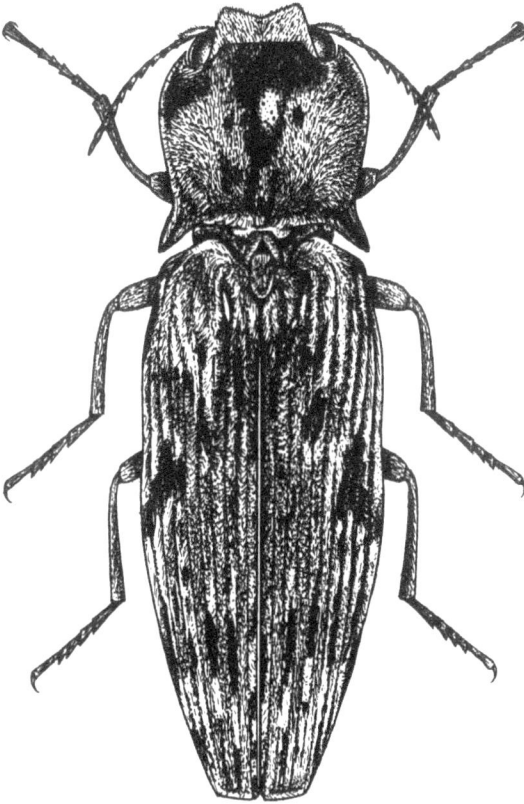

Figure 14 *Alaus* sp. (Elateridae).

Family EUCNEMIDAE

This family is closely related to the click-beetles but the members do not have the faculty of snapping themselves into the air like the latter. The species are usually stocky, cylindrical and with head, prothorax and hind body compact and streamlined. The legs may fit closely to the body and by shaming death the beetle may appear like wood or rubbish. Most of the species are brown, black or mottled and inconspicuous. Body length is about 5–12 mm. The larva is pale, narrow and flattened, with small head and no legs. Most of the species bore in wood of damaged or dying trees. New Guinea genera include *Arganus, Dendrocharis, Drapetes, Dromoeolus, Fornax, Galba, Hylocares, Potargus.*

Superfamily Cantharoidea
Family LAMPYRIDAE (Fire-flies)

The members of the family are familiar to many because of the luminous hind portion of the abdomen. Possibly all New Guinea species produce light. The pronotum is arched anteriorly in outline as viewed from above, hiding the head as a rule. The antenna is usually somewhat serrate, and the legs are rather short. The elytra are fairly flat and somewhat leathery, and are usually separately rounded behind. The body is usually black or brown, and the prothorax is sometimes pale or red.

Individuals of this family are abdundant in some parts of New Guinea and neighboring islands, especially in the lowlands and in New Britain. One small species lives on tidal coral formations and the adults may be seen, at dusk, flying over the spray on the coast line around Madang. There is an interesting phenomenon of synchronous flashing of the luminous organ among a great number of individuals of one species at a particular locality, with variations on the timing, synchronization and other aspects, depending on location or species. Light is produced by low-temperature oxidation fostered by particular enzymes.

The lampyrid larvae are flattened, tapering and somewhat leathery. The larvae may also be luminous. There appear to be rather few species in New Guinea, but this is not a very large family. New Guinea area genera include *Atyphella, Lampyris, Luciola* (Fig. 15; Pl. 6f), *Photinus, Pterophanes, Pteroptyx* (Pl. 6e).

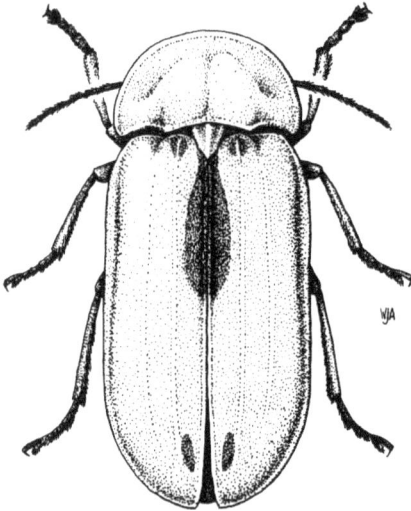

Figure 15 *Luciola obsoleta* Olivier (Lampyridae).

Family CANTHARIDAE (Soldier beetles)

This family shares the same superfamily, Cantharoidea, with the preceding family and the following one. All have in common somewhat soft, leathery, loosely fitting elytra, and bodies which are flattened, and more or less slender and parallel-sided, or slightly broadened behind. The Cantharidae usually have smooth and often slightly pubescent elytra. Most of the species are orange, straw-colored or black, often partly pale and partly

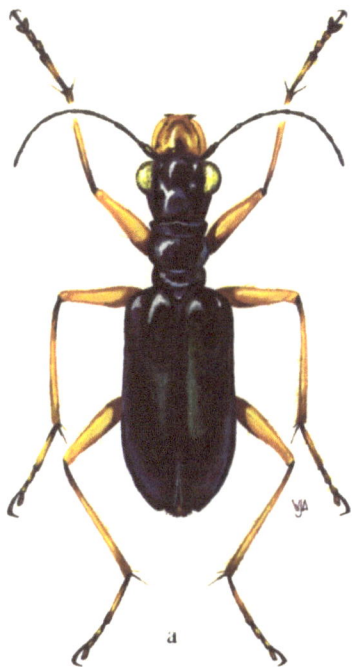

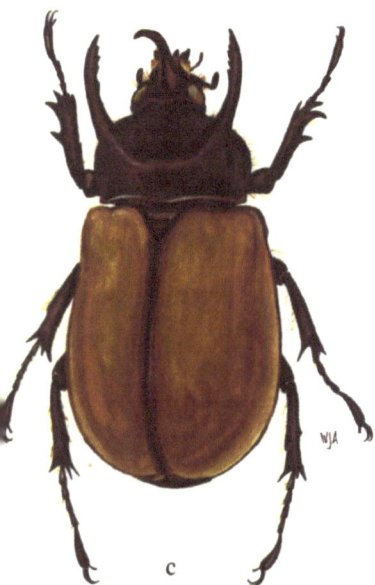

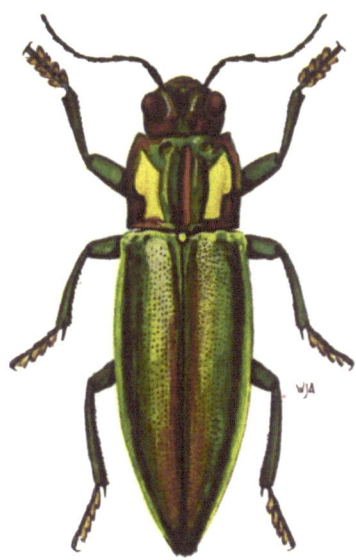

Plate 1

a. *Therates labiatus* Fabricius (Carabidae);   b. *Neolamprima adolphinae* Gestro (Lucanidae);
c. *Eupatorus beccarii* (Gestro) Scarabaeidae: d. *Cyphogastra* nr. *timorensis* (Buprestidae).
      Dynastinae);

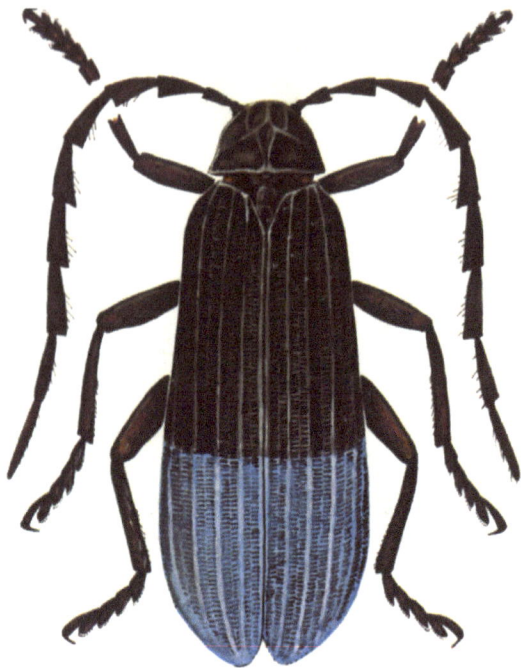

a

b

Plate 2

a. *Cladophorus* sp. (Lycidae);

b. *Eucymon bipustulatus* Gorham (Endomychidae);

c. *Bothricara pulchella* Guerin (Lagriidae);

d. *Lagriomorpha indigacea* Young (Anthicidae).

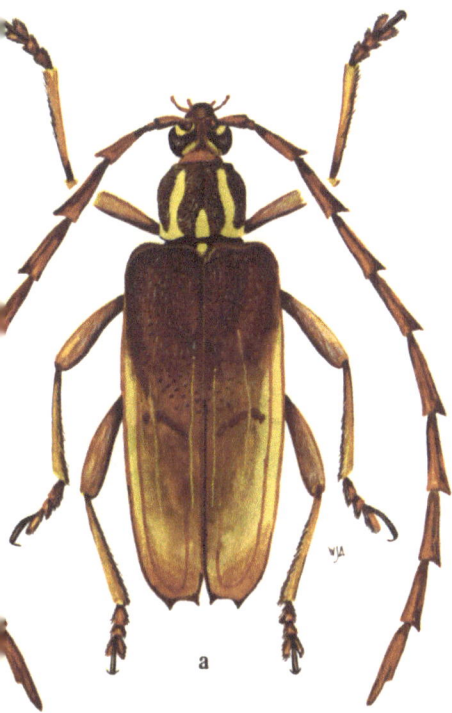

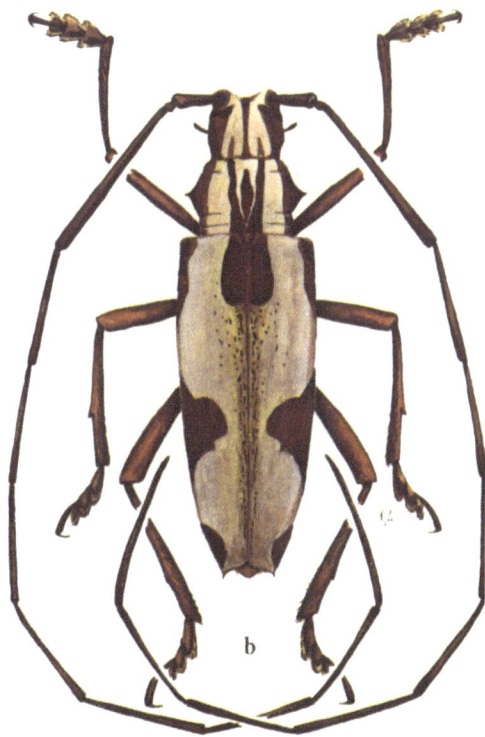

Plate 3

a. Coptopterus latus Gressitt (Cerambycidae:
   Cerambycinae;
c. Glenea lefebueri (Guerin)
   Cerambycidae: Lamiinae);

b. Epepeotes rarus (Thomson) (Cerambycidae:
   Lamiinae);
d. Promechus pulcher Gressitt & Hart
   (Chrysomelidae: Chrysomelinae).

a

b

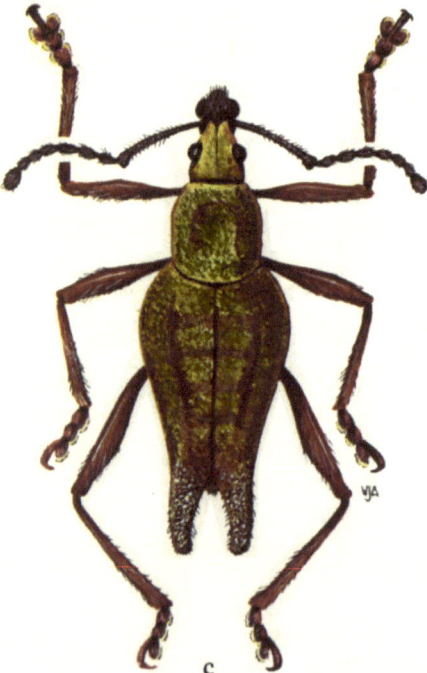

c

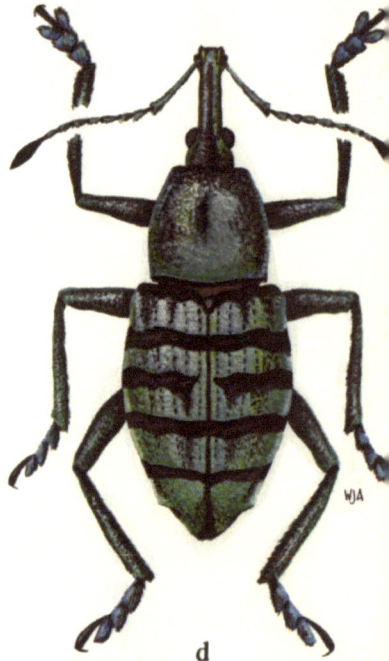

d

Plate 4

a. *Callistola puchra* Gressitt (Chrysomelidae: Hispinae);

b. *Aphorina australis* (Heller) (Apionidae)

c. *Hellerrhinus papuanus* (Heller) (Curculionidae: Otiorhynchinae):

d. *Eupholus geoffroyi* Guerin (Curculionidae: Leptopiinae).

black. Both prothorax and elytra are somewhat flattened, and the former is more or less rounded anteriorly, or rounded-oblong, usually broader than long. The antenna is slender and fairly long.

These beetles occur commonly on herbs and trees, often on the flowers. The larvae are active like the adults, and are often somewhat brightly colored and velvety. Both larvae and adults are predaceous upon small or soft-bodied insects. The adults of most of the species range between 6–20 mm long, but one species in the highlands, *Chauliognathus gibbosus* Wittmer (Pl. 6g) ranges up to 25 mm in length. It is one of the largest species in the world. It has a strong 2-lobed swelling on the prothorax, is black anteriorly and orange posteriorly, and is common at Daulo Pass and in the Upper Chimbu Valley. Other genera in the country are *Astychina, Sphaerarthrum, Neogressittia, Silvanotelus, Onychotelusia, Guineapolemius, Maltypus.*

Dr Walter Wittmer (1964 etc) has published a number of papers in recent years on the New Guinea members of this family, as well as the Melyridae, in *Pacific Insects, Nova Guinea* and other journals.

Family LYCIDAE

Members of this family are slender or slightly widened posteriorly, and range 5–25 mm long. They are usually easily recognizable by the ridged or reticulated prothorax and long, flat, finely net-veined elytra. The prothorax covers part of the head. The antenna is often serrate or flabellate, moderately long, and the legs are somewhat flattened. There are interesting colors among the New Guinea species. Many are red, orange and/or black, but some are unusual shades of pastel blue or green, often in combination with black or orange. Lycids are distasteful to birds, and this apparently accounts for various other insects having evolved body forms and patterns to mimic particular species, or common types, of lycids. Mimics seem not to be so evident in New Guinea as in Australia, C. America, etc. However, this family is extremely well represented in New Guinea, with a probable few hundred species. They are mostly montane, in altitude zones III and IV.

The larvae of the lycids are somewhat flattened, elongate and fairly active. They are not velvety like cantharid larvae. They lack caudal append- ages. Both larvae and adults are predaceous. Sometimes the adults feed on adult beetles, and vice versa. Among New Guinea genera are *Calochromus, Cladophorus* (Pl. 2a; Pl. 6h, i), *Leptotrichalus, *Malacolycus, Metriorhyn- chus, Mimotrichallis, Plateros, Trichalus, *Xylobanomorphus, Xylobanus.*

Superfamily Dermestoidea
Family DERMESTIDAE

This group is important in temperate areas but is poorly represented in

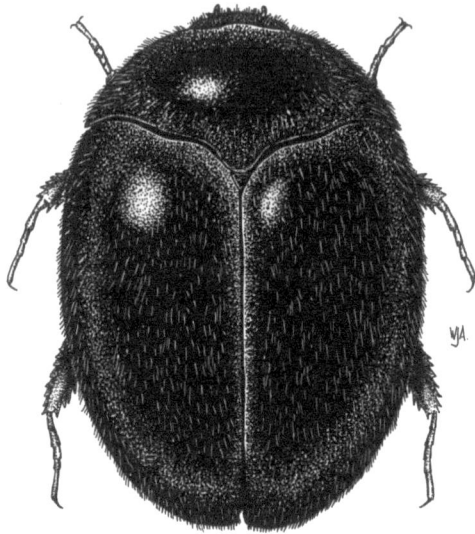

Figure 16 *Attagenus undulatus* (Motschulsky) (Dermestidae).

New Guinea, though occurring in imported dried food products. The species are mostly small and black or brown, mostly ranging from 2–7 mm in length. The majority of the species are scavengers, feeding on dried foods, hides, dead insects or in birds nests. Adults of some species visit flowers. The larvae are relatively short, and are rather bristly. New Guinea genera include *Anthrenus, Attagenus* (Fig. 16), *Dermestes*.

Superfamily Bostrychoidea
Family BOSTRYCHIDAE

Members of this family are borers in dead wood. They are mostly black, cylindrical, 6–10 mm long, with short antenna bearing three flattened and widened terminal segments. The posterior end of the elytra are usually more or less vertical, truncated and sculptured. They are similar in shape and in habits to some bark beetles, but usually bore in older and drier wood, such as posts or even furniture. Larvae are soft-bodied, tapered and curved, and are inside the wood. New Guinea genera include *Bostrychopsis, Dinoderus, Heterobostrychus, Xylothrips* (Fig. 17).

The family PTINIDAE includes short-bodied species feeding on dried foods.

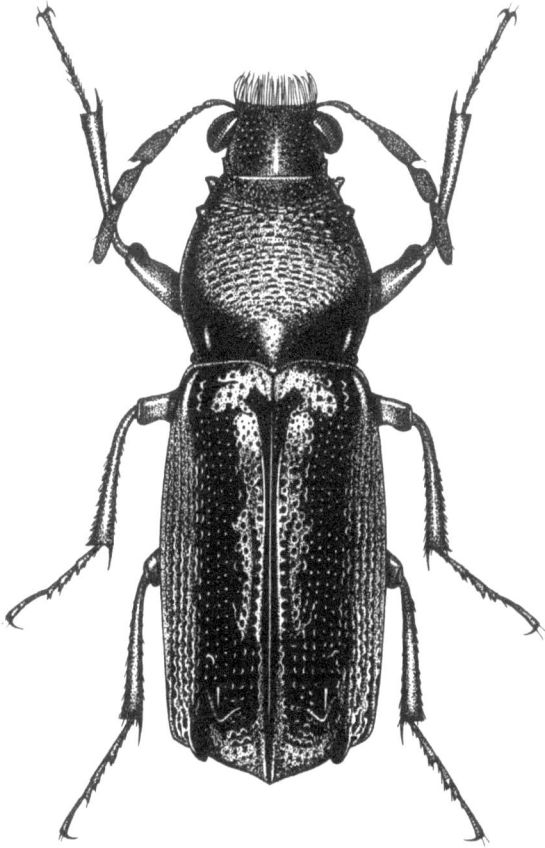

Figure 17 *Xylothrips religiosus* (Boisduval) (Bostrychidae).

Superfamily Cleroidea
Family CLERIDAE

These are usually small beetles 5–12 mm in length. The general body outline is superficially rather like the tiger beetles (Carabidae: Cicindelinne). They are parallel-sided insects with long legs and large downwardly directed mandibles. Unlike the tiger beetles, the elytra, legs and head are usually covered with short or long fine hairs and bristles, and the antenna is widened towards its termination. Most of the New Guinea species live on logs or the trunks of standing timber, and often large numbers may be

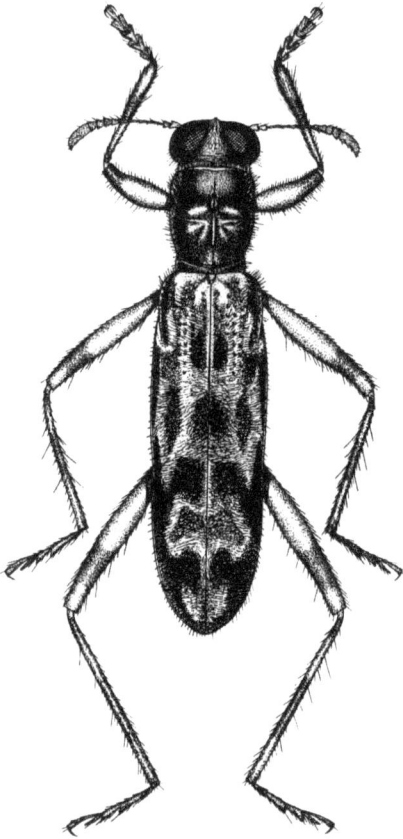

observed on the trunk of a dying tree which is infested with Platypodinae or Scolytinae and other borers. A few minutes observation will reveal that the Cleridae are effective predators, running down and capturing many of these smaller beetles. Flower-frequenting species are uncommon in New Guinea. Most of the species are prettily marked in complex patterns of shades of green, gray and black. There are some which live on the under surfaces of logs lying just above the ground, and the color pattern in these is much darker green or brown. A very large brown clerid at least 30 mm in length is rarely found in the tunnels made by cerambycid larvae in the interior of tree trunks.

Necrobia rufipes De Geer is the familiar copra beetle with bright shiny blue elytra and red head and prothorax, some 5–6 mm in length. It sometimes swarms among old copra.

Figure 18 *Omadius semicarinatus* (?) **(Cleridae).**

Other genera occurring in New Guinea are *Omadius* (Fig. 18), *Stigmatium, Tenerus, Cylidrus* and *Tillus.*

Family MELYRIDAE

This family, also sometimes called Malachiidae, is similar to the Canthari-dae, but is not as closely related to it as was indicated in older texts. Mem-bers are often shorter-bodied, setose and most of the New Guinea species have short elytra. Some, however, are very slender with short elytra, so may be confused with Staphylinidae. The elytra of melyrids are usually each about three times as long as broad, as opposed to those of staphylinids which are about as long as broad, more or less. The hind wings are not folded in such an intricate manner as in the Staphylinidae. Many of the New Guinea members of Mely-ridae are 4–9 mm long and dark steel blue, or red, brown and/or black. These beetles are active, often seen on the wing, and frequent leaves and flowers.

Melyrid larvae resemble those of clerids, and this family has other similarities with the Cleridae. The larva has a pair of caudal appendages. The mely-rids are predaceous both as larvae and adults, as with the clerids, as well as with the cantharids. The adults often more nearly resemble cantha-rids than clerids, in superficial appearance. Among New Guinea genera are *Attalus, Carphurus* (Fig. 19; Pl. 7b), **Neocarphurus, Paracarphurus, Scelocarphurus.*

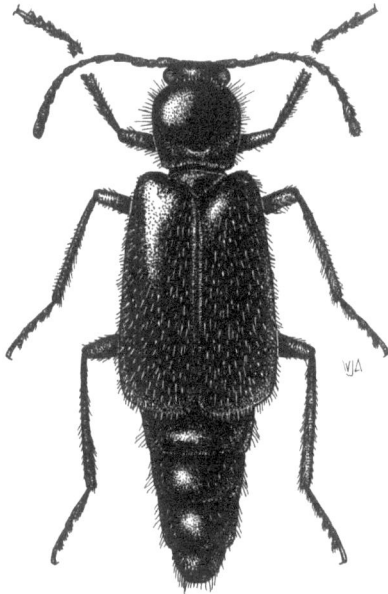

Figure 19 Family Melyridae.

45

Superfamily Cucujoidea
Family NITIDULIDAE (Souring beetles)

This family contains small oval beetles 1–8 mm in length. They tend to be flattened and the antenna is short but has a conspicuous widening in the three terminal segments, to produce a small club. The elytra are usually somewhat shortened to expose the terminal abdominal seggment. The adult insects and their larvae frequent rotting fruit and fungi, although a few species congregate on the exposed oozing sap of the stumps of recently felled trees, or live under moist bark, in dried fruits or in flowers. The larvae resemble some fly maggots and are slightly flattened.

Genera known from New Guinea include *Haptonchus*, *Megauchenia*, *Ithyphenes*, *Brachypeplus*, *Cryptarcha* and **Adocimus*. *Carpophilus dimidiatus* is a pest of stored grain (Szent-Ivany, 195b). Other Carpophilus (Fig. 20) and *Urophorus* comprise cosmopolitan species. *Lasiodactylum* attacks *Syzygium* (Szent-Ivany & Catley, 1960). Also Pl. 6d.

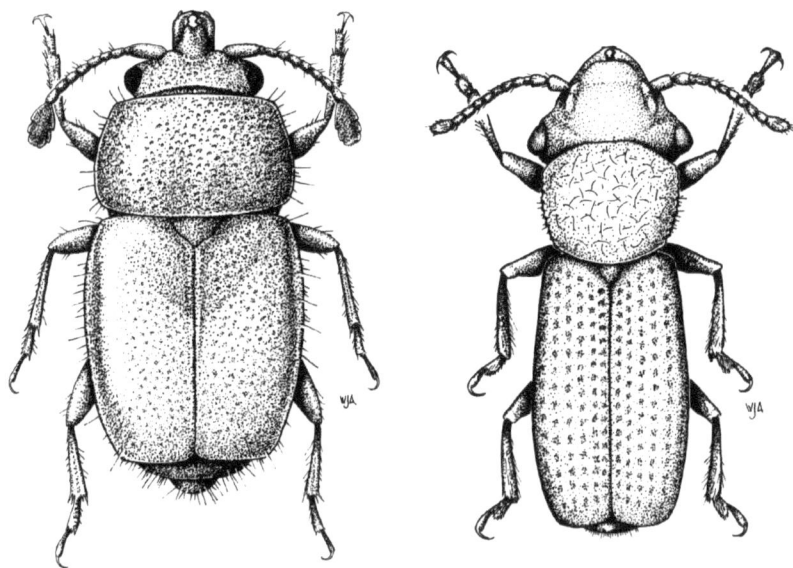

Figure 20 *?Carpophilus* sp. (Nitidulidae). **Figure 21** *Mimemodes* sp. (Rhizophagidae).

Family RHIZOPHAGIDAE

Small flattened beetles with large triangular heads concave above. Elytra slightly abbreviated. Species mostly brown with minute hairs in rows on elytra. Tarsi 5-5-5 in ♀; 4-4-4 in ♂. Length 2–3 mm. Some temperate species prey upon bark beetles. The species illustrated (*Mimemodes sp.;* Fig. 21) was collected on sunflower seed heads.

Family CUCUJIDAE

This family comprises very flat beetles which live under tree bark, as a rule. A common type is very flat and slender, and has long slender antenna and fine longitudinal ridges on the elytra. These are usually brown in color and 8–15 mm in length (Fig. 22). Some species in Asia are bright red or blue. The larvae are dirty whitish and are also flattened and live under bark. They may be scavengers, and some feed in stored foods. Genera include *Ancistria, Hectarthrum* and *Heliota.*

PASSANDRINAE – This subfamily has often been treated as a separate family Passandridae. A New Guinea species *Hectarthrum trigeminum*

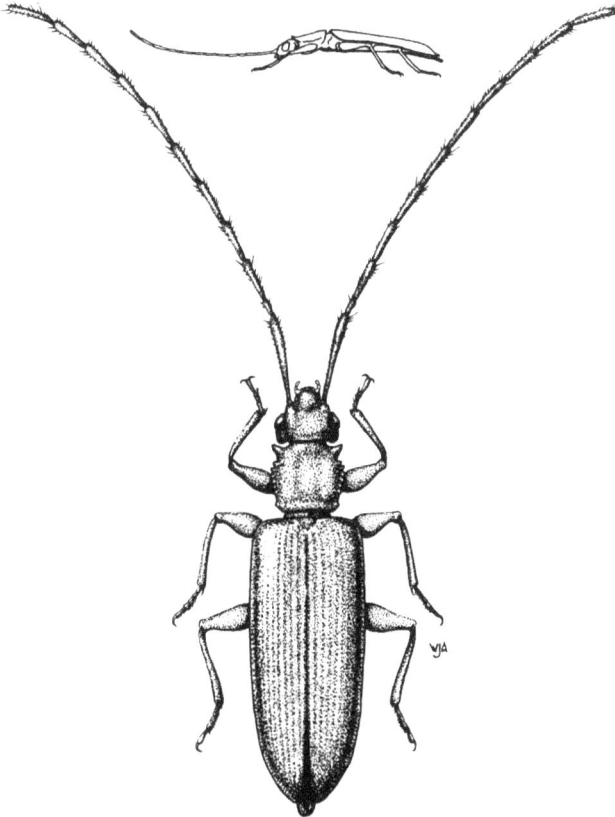

Figure 22 *Heliota* sp. (Cucujidae).

Newman (Pl. 7a) is rather conspicuous, being 20 mm long, moderately broad and very shiny black, with bead-like antenna and a few deep grooves on the elytra.

Family LANGURIIDAE

This family is closely related to the Erotylidae. There are not very many species in this group, but they are of moderate size and some of them are often encountered. The species are all slender and nearly cylindrical, as well as fairly smooth. Most of the species are red and black or red and blue, with the prothorax often red and the elytra usually dark. The antenna is fairly short and has a flattened club. Most species are 7–12 mm long. The genus *Coenolanguria* (Fig. 23) occurs in New Guinea.

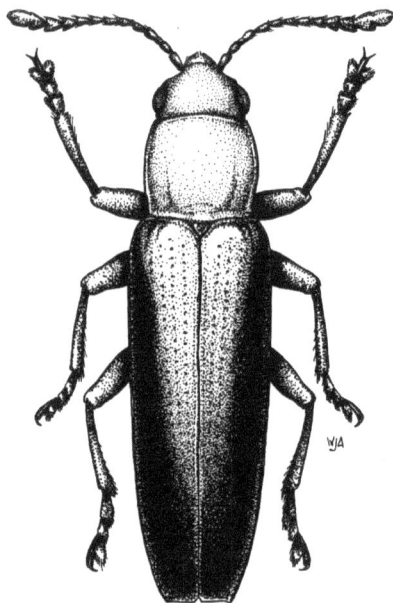

Family EROTYLIDAE

Members of this family are sometimes called the 'pleasing fungus beetles' and are often handsome. They are mostly narrowly elliptical in dorsal outline and fairly deep-bodied, being strongly convex above and well sclerotized. The antenna is moderately short and has a flattened club. The legs are short. The body surface is smooth and often shiny. Colors vary from black to red and yellow, often with bands or spots (Pl. 7c). Size varies from 5–18 mm in length, with the breadth one-fourth or one-third of the length.

Figure 23 *Coenolanguria papuensis* (Crotch) (Languriidae).

The larvae are pale, spiny and bear a posterior pair of appendages. Both larvae and adults feed in fungi or in fungus-infested wood. The adults are often found on the undersides of dead logs, where fungi are growing. Genera in New Guinea include *Aulacochilus, Encaustes, Episcaphula, Plagiopisthen, Camptotritoma, Simocoptengis* (Pl. 7c), *Sphaerotritoma, Spondotriplax.*

The related family CISIDAE or Ciidae comprises quite small cylindrical beetles which live in fungi or fungus-infested wood. They are shaped like bark beetles, but have the three segments of the antennal club quite separate.

COCCINELLIDAE (Ladybird beetles)

New Guinea has a very large number of species of native ladybird beetles. Although the majority of the species are small and uniformly dark brown or black in color, some have conspicuous yellow and black or red and black color patterns resembling the familiar ladybirds of the temperate regions of Europe and America.

The larvae are free-living and active with thoracic legs and a rather long attenuated abdomen. They are usually spotted black or gray on a whitish ground color, but some have quite bright red or yellow markings. The pupa may be seen attached to the underside of leaves in areas where adults and larvae are common. Both larvae and adults are well known through their role in controlling aphids and scale insects of which they consume large numbers. The prey of most of the New Guinea species is, however, in-

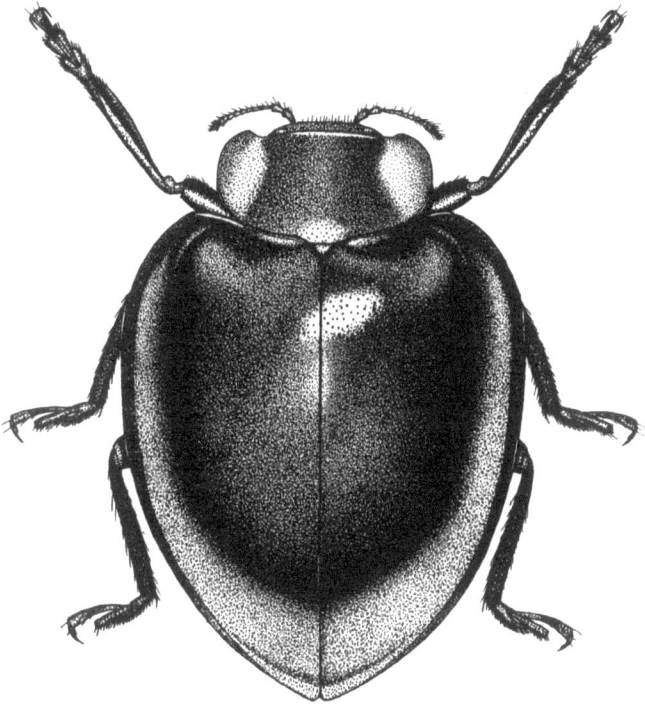

Figure 24 *Epilachna haemorrhoa* Boisduval (Coccinellidae).

completely known and much needs to be learned of their biology.

Among the genera recorded from New Guinea are *Coelophora, Ileis, Harmonia, Cheilomenes, Verania, Anisolema* (Pl. 7d), *Coccinella, Neda* (Pl. 7e), **Karubakana, Scymnus, Gyrocaria, Synia, Cryptolaemus* and *Epilachna* (Fig. 24; Pl. 7f). The species of *Epilachna* (Epilachninae) are not insectivorous like the others, but are plant-feeding and some of them consume leaves of beans and some other vegetables.

Family ENDOMYCHIDAE

This family is related to the preceding one and its members average slightly larger and are flatter and less evenly rounded or oval and less evenly convex. Many of them are bluish in color with some reddish markings. Their antennae are somewhat clubbed and are much more conspicuous than those of the ladybirds. The larvae are blackish and bristly and actively move around on dead logs and dead branches as do the adults. Some of them are fungus-feeders. They frequently occur in considerable numbers. Several species belong to *Encymon* (Pl. 2b).

Family COLYDIIDAE

This family comprises mainly small beetles 3–9 mm in length, of varied form. Most of them are dull in color and many are oblong or slender and often somewhat flattened. A few species may be mottled or marked with pale, and one, *Dryptops* (Samuelson 1966) often has algae and fungi, or lichens, growing on the upper surface of the body (Pl. 7g).

The larvae are usually subcylindrical and have caudal appendages. Some members of the family are apparently wood-borers but they may feed upon fungi. This is not a particularly small family, but many of the species are obscure and small. Other genera in New Guinea include *Asosylus, Lapethus, Paraxiocerylon, Petalophora, Sosylus* and *Teredolaemus*.

Family LAGRIIDAE

Members of this family have a characteristic shape, being somewhat flattened and broad, but with a much narrower, somewhat cylindrical prothorax and fairly small head. The antenna is of moderate length, and is weakly thickened distally. Most of the species are dull gray, black or brown, but some have pale pubescent bands, and some are narrower and more shiny than most.

The larvae are cylindrical, pigmented and often hairy. They occur primarily in rotting wood or other rotting vegetable matter. Not much is known about the feeding habits of the adults, in spite of the fact that they are fairly common around logs and on vegetation. They may be scavengers or leaf-feeders but do not appear to eat very much. They appear to crawl a great deal and to fly rather little, as compared with cantharids which are

50

often found in the same situations. Genera in New Guinea include *Biro-lagria, Bothricara* (Pl. 2b), *Casnonidea, Lagria* and *Nemostira.*

Family TENEBRIONIDAE

This is one of the larger families of beetles, often called the 'darkling beetles.' They are of varied form, but mostly oblong or oval. The majority are black, either dull or shiny, and others are metallic blue or purplish.

Most of the Tenebrionidae are probably scavengers. Many of them live somewhat hidden in the leaf-mold, under rotting logs or in quite old logs and in cavities in bark, or under bark. Most of them are rather slow moving, but the longer-legged ones may be active when disturbed. This family is not very well known for Papua New Guinea as yet, but is now under intensive study by Dr Z. Kaszab of Budapest.

The larvae of Tenebrionidae are mostly slender, cylindrical and brown, being rather heavily sclerotized. The legs are short. The larvae occur mainly in rotten logs. They are easy to confuse with elaterid (click-beetle) larvae.

Dr Kaszab kindly supplied an outline of the New Guinea tribes and genera, which is abbreviated below for lack of sufficient space.

Pedinini: *Mesomorphus, Diphyrrynchus.*

Opatrini: *Gonocephalum.* Dull black, flattened and oblong; on ground.

Boleophagini: *Bradymerus* (Pl. 7h), *Scotoderes, Byrsax.* Prothorax with flattened sides; elytra with deep grooves, tubercles and/or punctures.

Dysantini: *Orcopagia.*

Diaperini: **Louverensia, Platydema, Ceropria.* Parallel-sided, rounded behind; smooth to slightly grooved.

Gnathidiini: **Szentivanya, Menimus.* More or less ovate; antenna of 10 instead of 11 segments.

Leiochrini: *Leiochrinus, Derispia, Leiochrodes.* Short, rounded; black or metallic; on logs.

Phrenapatini: **Pseudophthora.*

Ulomini: *Lyphia, Achthosus* (Pl. 7i), *Uloma, Sciophagus, Hypophloeus.* Oblong, black; elytra ribbed; legs short.

Helaeini: *Encara* (Pl. 8c), **Euhelaeus* (Pl. 8b).

Eutelini: **Tabarus* (Pl. 8d).

Tenebrionini: **Lomocnemis, Setenis* (Pl. 8a), **Graptopezus, Encyalesthus, Anthracias.* Usually oblong, often large; elytra usually grooved.

Lypropini: *Lyprops* (Fig. 25), *Pseudolyprops.*

Cnodalonini: *Phenus, Apterophenus, *Microphenus, *Cerandrosus, *Cataphanus, *Pezophenus, Agymnonyx, *Neotheca, Thesilea, Cariotheca.* Elliptical oblong; elytra seriate-punctate; black to metallic.

Amarygmini: *Platolenes, Amarygmus, *Spathulipezus.* Convex, ovate; legs long; antennae slender; black to metallic.

Strongyliini: *Lophocnemis, *Parastrongylium, *Heterostrongylium, Strongylium.* Slender; antennae and legs long and slender; elytra usually seriate-punctate.

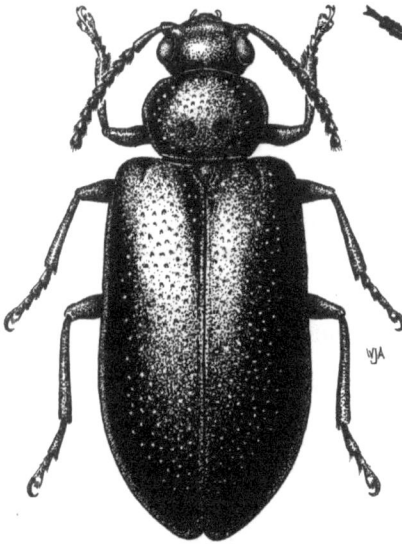

Figure 25 *Lyprops atronitens* **Fairmaire (Tenebrionidae: Lypropini).**

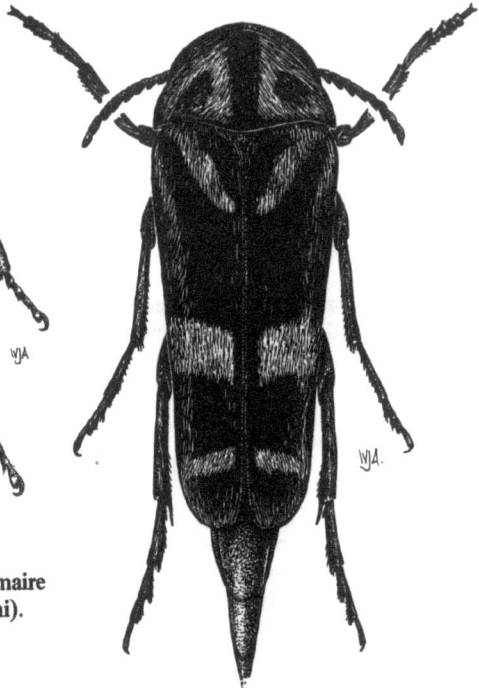

Figure 26 *Glipa mixta* **Fabricius (Mordellidae).**

Family MORDELLIDAE

These are small beetles which are characteristically hump-backed, having the body arched back in a convex curve from the head to the termination of the abdomen. This effect is exaggerated by the development of the end of the last segment of the abdomen into a spinous process. The process assists the beetle to jump and hop when it is pursued. The beetles are usually black, sometimes with white stripes or spots. One species 8–10 mm in length is brown with a complex pattern of orange and white markings, and it may be commonly encountered resting on leaves in hot sunshine in the lowlands. It flies off very rapidly when disturbed. All the beetles are attracted to flowers where they may sometimes be seen in large numbers. The larvae are predaceous, parasitic or wood-boring. Little is known of their habits in New Guinea.

Genera recorded from New Guinea include *Mordella, Glipa* (Fig. 26), *Ermischiella* and *Mordellistena*.

Family RHIPIPHORIDAE

This family is related to the Mordellidae, and some of its members have a similar flea-like body form but are less tapered like a spine posteriorly. Others are flatter and less arched above. All have a pectinate or flabellate antenna, at least in the male. Some have reduced elytra. The size of these beetles is in the range of about 6–10 mm in length. The colors are often brown or black, but some have pale markings. The larvae in most cases are parasitic in bodies or egg-cases of certain insects, or nests of bees. The

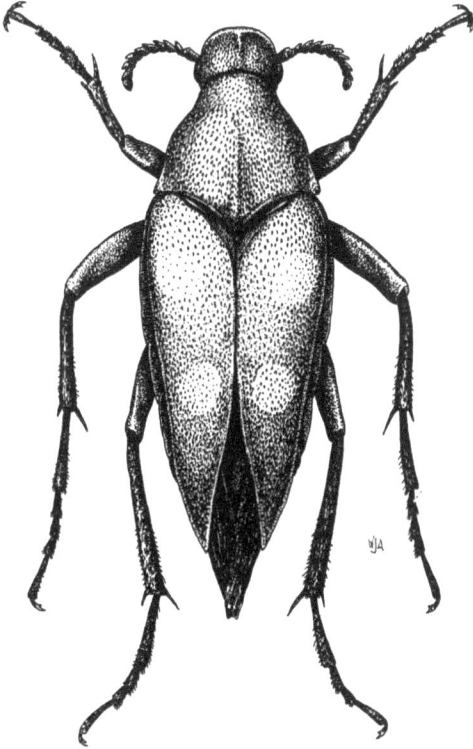

Figure 27 *Macrosiagon cucullatus* **Macleay (Rhipiphoridae).**

young larvae are bristly and very active, or wait for their hosts on flowers. After being carried to the nests by the bees or other insects, the larvae molt to a quite different-appearing legless grub and feed as parasites These beetles are in general scarce in New Guinea. Representatives in clude species of the genera *Falsorhipidius, Macrosiagon* (Fig. 27) and *Pelecotomoides* (Pl. 8e).

The family ALLECULIDAE includes oval species resembling some tenebrionids.

The family MELOIDAE, the blister beetles, is more poorly represented in New Guinea than in most parts of the world. These are medium-large beetles from black to brightly colored, some of which cause blisters on contact with tender skin. They are usually broader than members of the next family, some of which also cause blisters.

Family OEDEMERIDAE

This family is commonly present in the lowlands, the members sometimes being very abundant, but is not represented by many species. The beetles are slender, cylindrical or slightly tapering and with fairly long antenna and legs, and somewhat flexible and loosely fitting elytra. Colors are mostly yellowish brown to brown or pitchy and the beetles are mostly 8—20 mm long. The species are particularly abundant in coastal areas like coconut plantations and these are often the dominant beetles on atolls and small offshore islets. The larvae principally breed in rubbish, driftwood or in posts on which dogs urinate. The adults of some species have a body fluid or secretion which is irritating to tender skin. Blisters may be caused if a beetle is crushed under the clothing. The beetles are attracted to light and thus frequently enter houses. They are also attracted to flowers, such as coconut blossoms. If ingested with coconut drinks the result may be kidney inflammation. Members of this family may be confused with Canthari-dae, but have the prothorax sub-cylindrical instead of flat above, and have the tarsi 5-5-4. New Guinea genera include *Ananca* (Fig. 28), *Falsosessinia, *Mil-neum* and *Sessinia.*

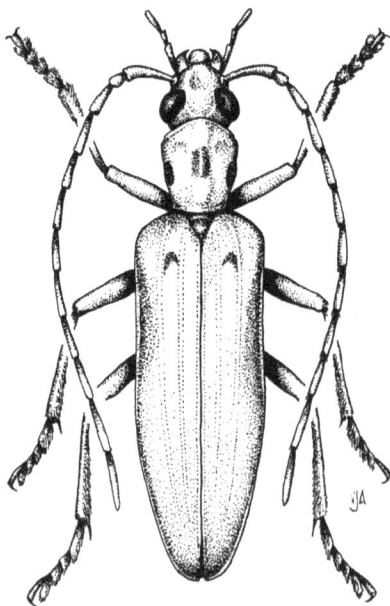

Figure 28 *Ananca kanak* Fairmaire (Oedemeridae).

Family ANTHICIDAE (Ant-like flower beetles)

This is a family of mostly quite small slender beetles, 2–9 mm long. They are usually easily recognizable by a longish slender neck, with the head often bent downward. Some species have a forward projection on the prothorax. Colors are largely dull brown or tawny, but may be darker or paler and banded or spotted. The larvae are slender, pale and slightly hairy. Genera in New Guinea include *Anthicomorphus, Tomoderus* and *Lagriomorpha*. Most species are probably scavengers, but little is known of the habits of New Guinea species (Fig. 29). *Lagriomorpha* (Pl. 2d) was wrongly placed in the family Pyrochroidae.

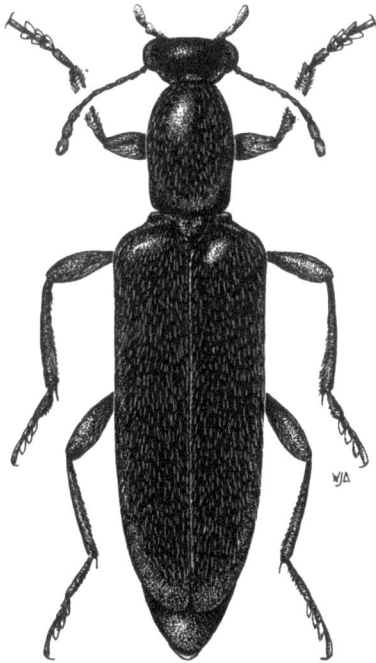

Superfamily Chrysomeloidea
Family   CERAMBYCIDAE (longicorns)

This large family includes a probable few thousand species in New Guinea and nearby islands. Among these are some of the

Figure 29 Family Anthicidae.

largest and most striking of beetles found anywhere. There is a great range in size, say 3–100 mm in length of body, and great variety in color, pattern and form. In general, members of the long-horned beetle family are elongate to quite slender, with antenna often two or three times as long as the body in males.

The larvae of these beetles are almost entirely wood-borers, feeding in the wood, or cambium, of living trees, dead branches, standing dead trees, or more rarely of fallen logs. The larvae are usually long and slender, slightly tapering posteriorly, and white or creamy, except for the brown to blackish head and partly brown pronotum. They may tunnel deep in the heart wood, or bore largely under the bark. These larvae are quite edible. They make a pupal cell inside the wood, after having nearly chewed through the bark. They then pupate in the cell, and the adult emerges through the bark after a period of hardening of the exoskeleton.

This family comprises several subfamilies, one or two of which are not known from New Guinea and nearby areas. The former subfamily Disteniinae is now placed as a separate family, although long considered within the Cerambycidae. Members of the DISTENIIDAE are extremely slender, with very slender antenna bearing an internal groove with long fine fringe which is often hidden in the groove. The prothorax is usually tuberculate at side.

PRIONINAE — This subfamily may be the most primitive one here and is fairly well represented in New Guinea. It includes mostly large to very large species, usually broad-bodied, and brown or black, with often two or three teeth, or more, on side of pronotum, or at least a flattened marginal extension. Many of the species bore in dead logs or roots. The adults are night-fliers and are seldom seen in the day-time, but they may be attracted to lights. Important genera here are *Xixuthrus, Olethrius, Osphryon* and *Agrianome* (Pl. 8g).

PARANDRINAE — This small group has been considered the most primitive in the family, but Crowson believes that it is specialized and derived from the Prioninae. The genus *Paranda* is associated with *Araucaria* trees in New Guinea. Its species are pale reddish brown, parallel-sided, flattened and of moderate size (12—18 mm long).

LEPTURINAE — This subfamily is largely a north-temperate group, completely lacking in Australia. It is represented in New Guinea by *Papuleptura* of small, relatively short dull species at high altitudes, and by *Elacomia,* longer tapering species banded with dark and pale, on islands at the far west end of the island of New Guinea. Body length of known species is 4—14 mm. Nothing is known of the habits of the New Guinea species. Larvae of many temperate species are borers in conifers, and the adults are usually seen on flowers.

CERAMBYCINAE — This subfamily is numerous in species in New Guinea, but is not the dominant group of the family as it is in Australia. This emphasizes the fundamental difference between the Australian and Papuan (zoogeographic sense) insect faunas in spite of the fact that there is considerable overlap, with species in common between the Cape York Peninsula and southern New Guinea. Members of this subfamily are diverse in form. In general, species in this subfamily are slender, more or less cylindrical, have the prothorax cylindrical or moderately swollen or sub-tuberculate at side, and are of varied patterns. The more primitive tribes have longer antennae, are often of larger size, and are mostly active at night, including flying to lights, while the more advanced tribes usually have shorter antennae, smaller bodies and are mostly active in the day time.

The antennae may be from about one-fourth as long as body to two or more times as long as body. Size may be 4—50 mm or more in body length. Color is highly varied with all manner of patterns. Body surface may be smooth or closely pubescent or punctured. Some species bore in living trees, or large semi-woody herbs, and many others bore in dead or dying branches or trunks. Some are fairly host-specific but not much precise data have been published on this subject for New Guinea.

Plate 5

a. *Dineutes (Merodineutes) archboldianus*
   Ochs (Gyrinidae);
c. Dytiscidae sp.;

e. *Plaesius cossyphus* Marseul (Histeridae);
g. *Priochirus* sp. (Staphylinidae);
i. *Phaeochrous emarginatus* Castelnau
   (Scarabaeidae: Hybosorinae).

b. *Hydaticus pacificus* Aube (Dytiscidae);

d. *Hydrophilus picicornis* Chevrolat
   (Hydrophilidae);
f. *Diamesus osculans* Vigors (Silphidae);
h. *Creophilus albertisi* Fauvel (Staphylinidae);

a. *Dineutes (Merodineutes) archboldianus*
   Ochs (Gyrinidae);
c. Dytiscidae sp.;

e. *Plaesius cossyphus* Marseul (Histeridae);
g. *Priochirus* sp. (Staphylinidae);
i. *Phaeochrous emarginatus* Castelnau
   (Scarabaeidae: Hybosorinae).

b. *Hydaticus pacificus* Aube (Dytiscidae);

d. *Hydrophilus picicornis* Chevrolat
   (Hydrophilidae);
f. *Diamesus osculans* Vigors (Silphidae);
h. *Creophilus albertisi* Fauvel (Staphylinidae);

Plate 6

a. *Alaus* sp. (Elateridae);

b. *Dermolepida noxium* Britton (Scarabaeidae: Melolonthinae);

c. *Papuana japenensis* Arrow (Scarabaeidae: Dynastinae);

d. Nitidulidae sp.;

e. *Pteroptyx cribellata* (Olivier) (Lampyridae);

f. *Luciola* sp. nr. *obsoleta* Olivier (Lampyridae)

g. *Chauliognathus gibbosus* Wittmer (Cantharidae);

h. *Cladophorus* sp. (Lycidae);

i. *Cladophorus?* sp. (Lycidae).

Plate 7

a. *Hectarthrum trigeminum* Newman (Cucujidae: Passandrinae);

b. *Carphurus viridipennis* Wittmer (Melyridae);

c. *Simocoptengis* sp. (Erotylidae);

d. *Anisolema* sp. (Coccinellidae);

e. *Neda fuerschi* Bielawski (Coccinellidae);

f. *Epilachna* sp. (Coccinellidae);

g. *Dryptops phytophorus* Samuelson (Colydiidae);

h. *Bradymerus nigerrimus* Gebien (Tenebrionidae: Boleophagini);

i. *Achthosus auriculatus* Gebien (Tenebrionidae: Ulomini).

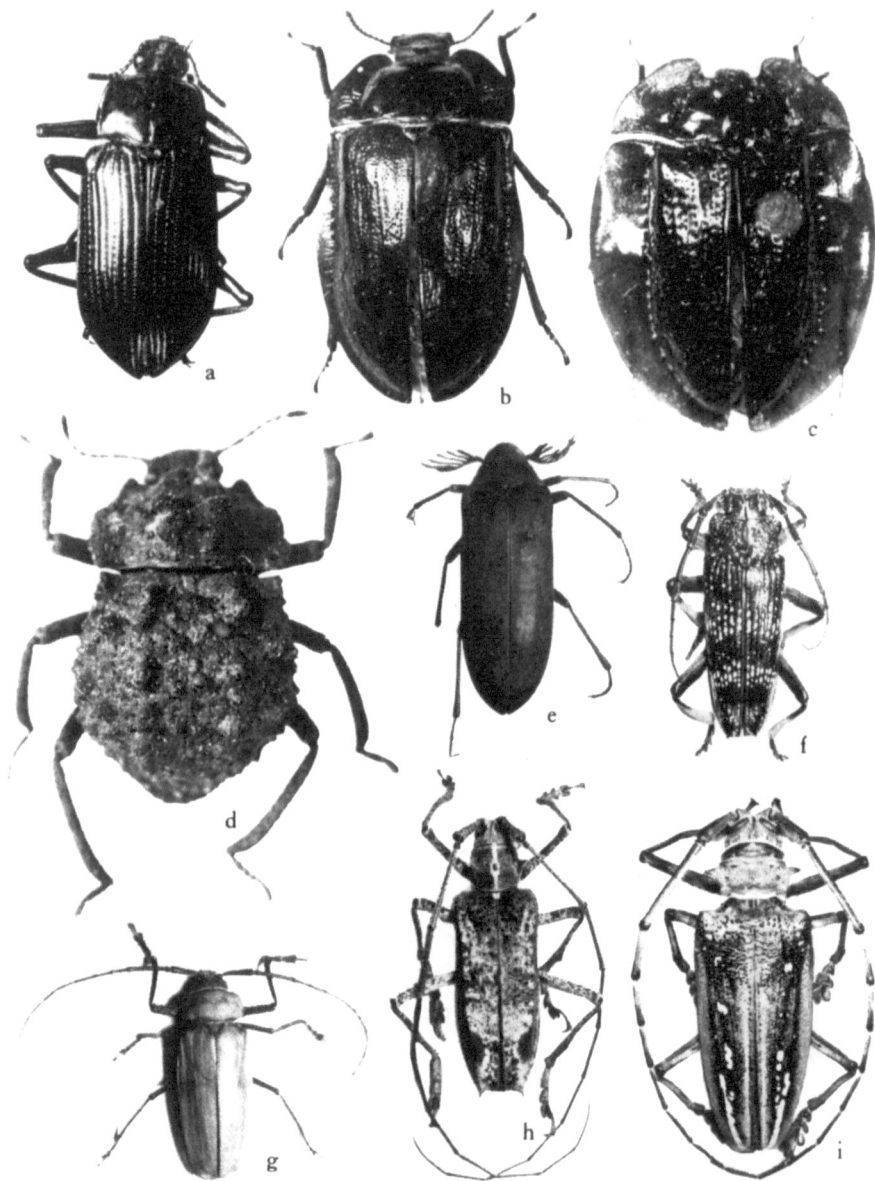

Plate 8

a. *Setenis illaesicollis* Fairmaire (Tenebrionidae: Tenebrionini);

b. *Euhelaeus speculiferus* Gebien (Tenebrionidae: Helaeini);

c. *Encara devidiens* Gebien (Tenebrionidae: Helaeini);

d. *Tabarus infernalis* Gebien (Tenebrionidae: Eutelini);

e. *Pelecotomoides* sp. (Rhipiphoridae);

f. *Sphingnotus insignis* Perroud (Cerambycidae: Lamiinae);

g. *Agrianome loriae* Gestro (Cerambycidae: Prioninae);

h. *Potemnemus* nr. *detzneri* Kriesche (Cerambycidae: Lamiinae);

i. *Batocera wallacei* Pascoe (Cerambycidae: Lamiinae).

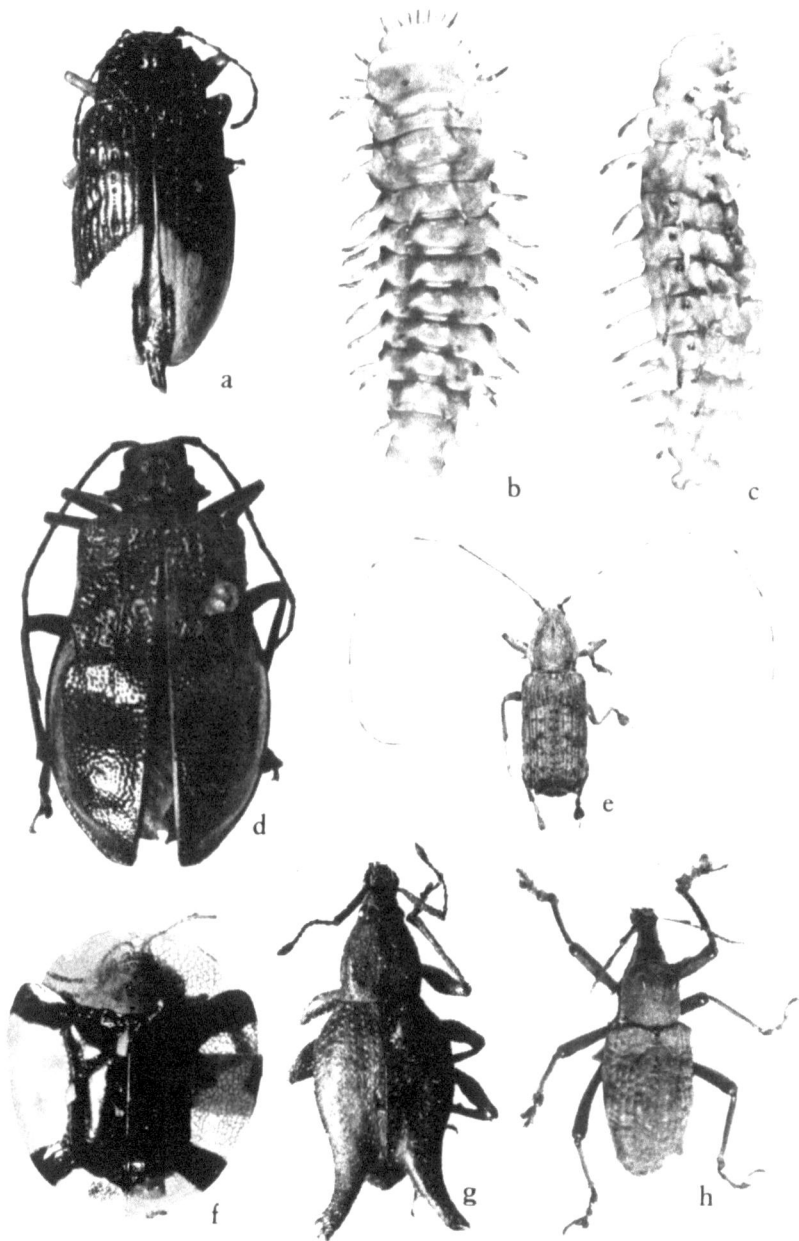

Plate 9

a. *Promechus pittospori* Gressitt & Hart (Chrysomelidae: Chrysomelinae);

b. *Promechus paniae* Gress. & Hart, larva, dorsal view;

c. *P. paniae,* larva, lateral view

d. *Gronovius imperialis* Jacoby (Chrysomelidae: Galerucinae);

e. *Ancylotrichis waterhousei* Jekel (Anthribidae);

f. *Aspidomorpha australasiae* Spaeth (Chrysomelidae: Cassidinae);

g. *Hellerhinus* sp. (Curculionidae: Otiorhynchinae);

h. *Gymnopholus nodosus* Gressitt (Curculionidae: Leptopiinae).

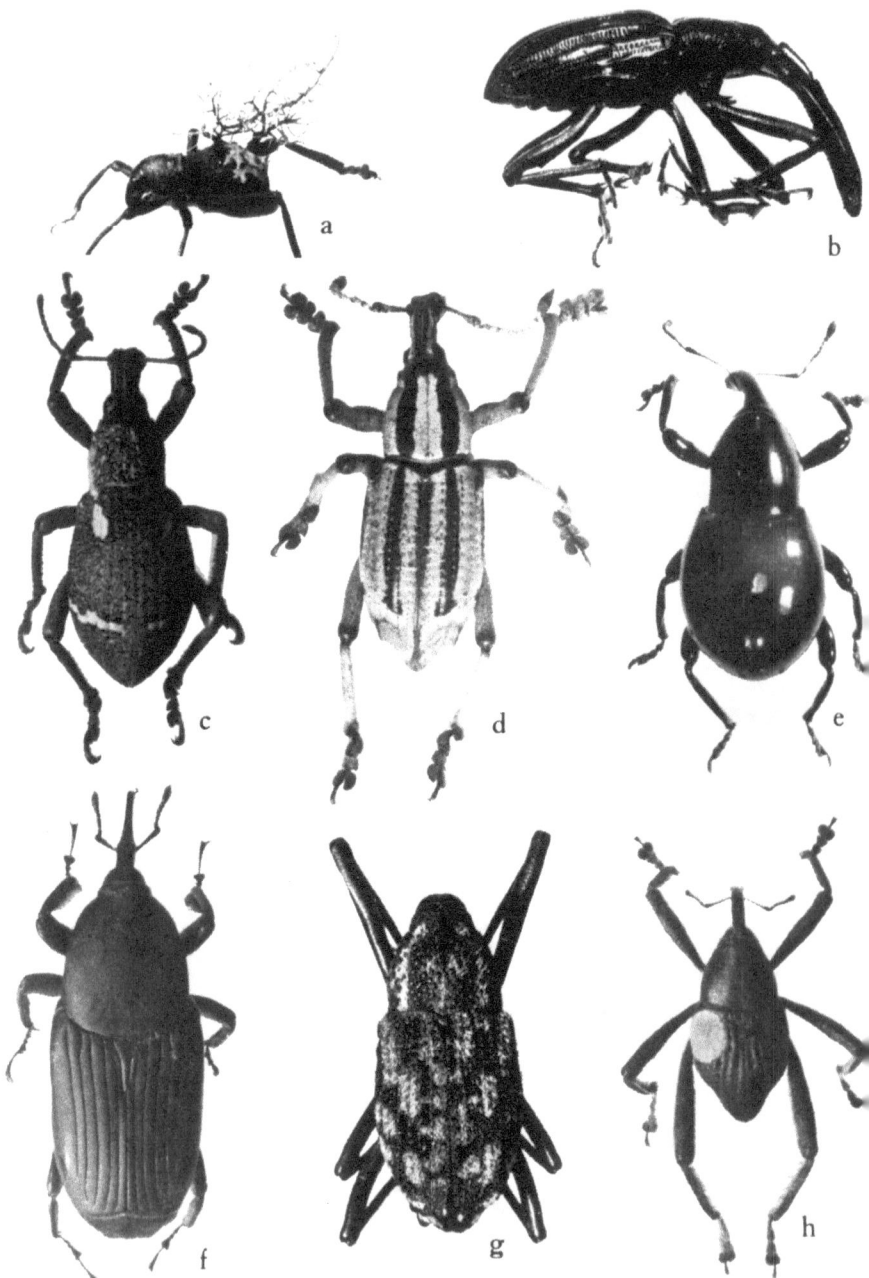

Plate 10

All Curculionidae:

a. *Pantorhytes lichenifer* Gressitt (Brachyderinae) with 2 kinds of lichen growing on elytra: *Usnea* above and *Parmelia* below;

b. *Vanapa oberthueri* Pouillaude (Hylobiinae);

c. *Rhinoscapha richteri* Faust (?) (Leptopiinae);

d. *Eupholus bennetti* Gestro (Leptopiinae);

e. *Carbonomassula cobaltina* Heller (Molytinae);

f. *Rhyncophorus bilineatus* Montrouzier (Rhyncophorinae);

g. *Disopirhinus* sp. (Cryptorhynchinae);

h. *Arachnopus* sp. (Zygopinae).

Twenty-three tribes with 76 genera are known for this subfamily from New Guinea. The more important tribes (with some of the largest genera) are Cerambycini (*Hoplocerambyx* zone I), Phoracanthini (*Coptocercus* II), Callidiopini (*Ceresium* II, *Tethionea* I), Piesarthrini (*Coptopterus* III), Calliprasonini (*Syllitus* I, II), Clytini (*Xylotrechus*, *Chlorophorus*, *Demonax* I–III).

LAMIINAE – This is by far the largest group of New Guinea longicorns, with a probable few thousand species. Members of this subfamily are usually characterized by a vertical front of the head, but a major complex of tribes (including Tmesisternini) has the head oblique like in the preceding subfamilies. Members of this subfamily usually have an oblique groove on the inner side of the fore tibia and another on the back side of the mid tibia. Size range for this subfamily is as for the family as a whole.

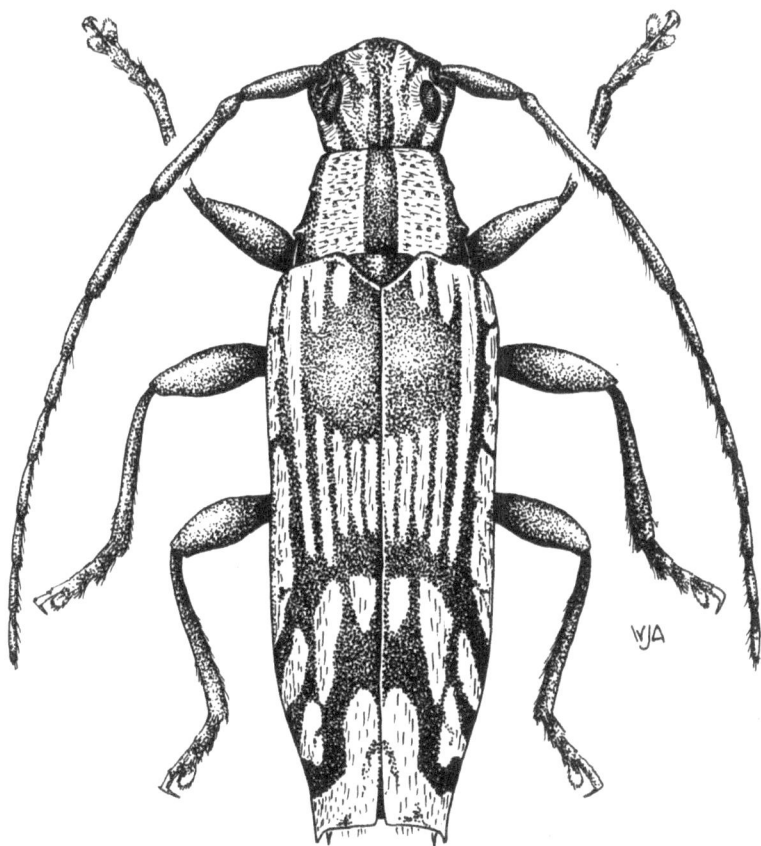

Figure 30 *Tmesisternus flavolineatus* Breuning (Cerambycidae: Lamiinae).

57

Members bore primarily in dead branches, trunks or logs, but some attack living trees, as cacao for example (some *Tmesisternus* and relatives), and some bore in large herbs. Among the latter are *Glenea elegans* (I) and relatives — often narrow, deep-bodied species of bluish black with conspicuous white spots. In general, the bulk of the species are useful in helping to break down dead wood in the forests, and coconut frond midribs in the copra plantations, for instance. This is all part of the natural recycling and soil formation of the native forests. The effect of these beetles on timber may be seriously damaging, as larger species make sizeable tunnels in the inner wood. An example is *Batocera wallacei* (I) (Pl. 8i), one of the largest-bodied beetles, which is 2 cm or more in diameter and up to 100 mm long, with very heavy antenna and legs. The claws of such large species are strong and sharp, and the mandibles are capable of drawing blood, being sharper than the much larger mandibles of most stag-beetles. This species and other *Batocera* bore in figs and related trees.

The life-cycles of most species probably involve only a few months at low altitudes, whereas in temperate countries they may occupy one, two or more years. While the larvae bore in twigs, branches, trunks or roots, the adults may feed on tender bark, flowers or juices from injuries. The eggs are usually laid in slits chewed into the bark.

The lamiines comprise a large number of tribes and an unknown number of genera, of which some of the more important are Lamiini (*Dihammus, Potemnemus* Pl. 8h, *Epepeotes* Pl. 3b, *Pelargoderus* I), Batocerini (*Batocera* I, *Rosenbergia* II), Mesosini (*Mesosa, Cacia* I), Gnomini (*Gnoma* I), Dorcaschematini (*Olenecamptus* I), Homonoeini (*Mulciber* I), Tmesisternini (*Pascoea, Sphingnotus* I, Pl. 8f, *Tmesisternus* I—IV Fig. 30), Niphonini (*Pterolophia* II, *Prosoplus* I), Apomecynini (*Apomecyna, Ropica* II), Ptericoptini (*Sybra, Meximia* I), Acanthocinini (*Exocentrus* II), Gleneini (*Glenea* I Pl. 3c), Phytoeciini (*Oberea* II).

Family BRUCHIDAE (Pea weevils)

The pea weevils are very short-bodied beetles with the elytra abbreviated to show a large pygidium. The legs are fairly short and the antenna is relatively long and somewhat serrate. Colors are usually dull brown, red-brown, tawny or dull mottled. Length of body is usually 3—5 mm. The larvae are pale and grub-like with short arched body, narrowed posteriorly. They feed inside of peas, beans or other seeds, primarily of legumes. Some feed in growing pods and some in dried legumes.

This family is poorly represented in New Guinea. Some cosmopolitan species, mostly in *Bruchus* (II) and *Acanthoscelides* (II; Fig. 31), may be imported in dried legumes.

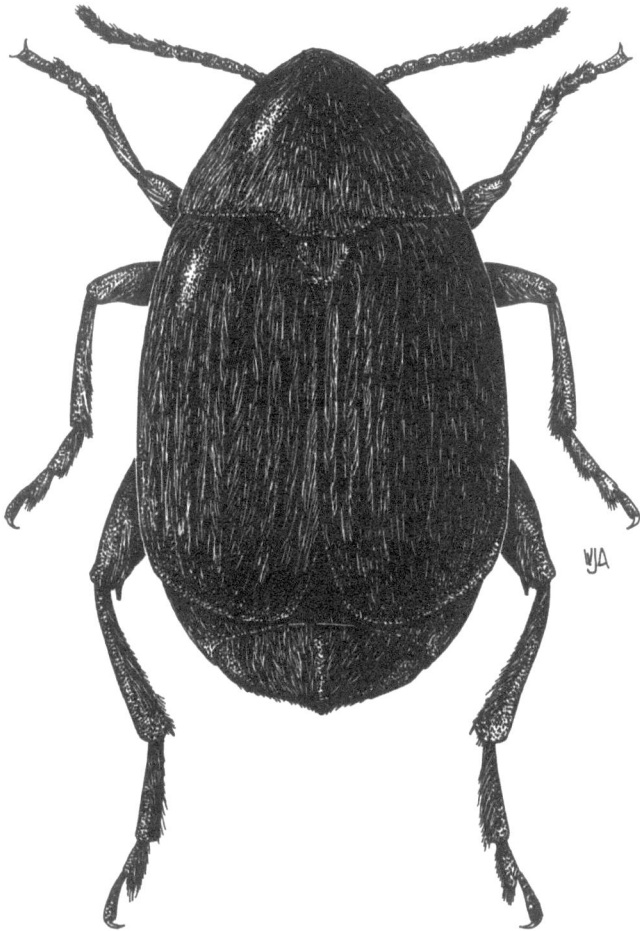

Figure 31 *Acanthoscelides obtectus* (Say) (Bruchidae).

Family CHRYSOMELIDAE (Leaf beetles)

This family is one of the largest among the beetles, perhaps second or third in numbers of species in the world. It is very well represented in New Guinea, by almost 5,000 species, of which only about 1,000 are named. The members of this family are almost entirely leaf-feeders in the adult stage. In the larval stage many are root-feeders, but others are leaf-feeders, leaf-miners or bark-feeders, with just a few living inside of semi-woody plants. Most of the members of this group are medium-small, but some are minute and some moderately large (1.2–30 mm). They are slender, ovate or nearly round in outline, usually somewhat flattened or moderately convex. Some are easily confused with lady-bird beetles, or with members of some other families. They can be distinguished from most beetles by all tarsi being similar, with four apparent tarsal segments (actually 5, but the 4th minute), and by the antenna usually shorter or much shorter than the body, but rather slender and not clubbed or geniculate, and the body usually fairly short.

The larvae are of several types in superficial appearance. Many are pale and naked, but some are bristly and dark, like some ladybird larvae. Others are more flattened, and some of these place their excreta on their backs for camouflage. In some of the latter cases the molted larval skins are retained on the caudal fork and the excreta are extruded onto the molted skins. Life cycles are usually of short duration, permitting several generations per year.

There are 16 subfamilies in this family, of which 12 are known from New Guinea. The first few subfamilies, generally considered the more primitive ones, are poorly represented here, but some of the more advanced ones are very well represented.

SAGRINAE — This subfamily includes large metallic species with strongly swollen hind femora. They occur only in the lowlands, and are not usually abundant. Their larvae are exceptional in being borers in woody bean vines. Probably only the genus *Sagra* occurs here.

DONACIINAE — These comprise slender beetles, flattish above, which live on water plants. The larvae are unusual in living under water throughout their development, obtaining air through the stems of water plants. This group is so far only known from the Lake Murray area on the Fly River, where it is represented by a species of *Donacia*. It is reddish brown tinged with bronzy, and about 7—10 mm long.

CRIOCERINAE — Members of this group are mostly subcylindrical but with the prothorax much narrower than the elytra and the former usually constricted before and behind middle. Most of the species are 4—10 mm long and are red, orange, blue or black, with usually two of these colors. The larvae are naked, swollen behind, with small heads, and feed on the surfaces of leaves of gingers and other monocots. Some of them feed on banana leaves. These include *Lema papuana*. Some species feed on orchids. Other genera present are *Lilioceris* and *Stethopachys*.

ORSODACNINAE (ZEUGOPHORINAE) — These are few in species, consisting of small beetles with somewhat swollen femora and weak jumping habit. They have only been found in the highlands, and nothing is known of the habits of the local species. The known species belong to *Zeugophora*.

CLYTRINAE — This subfamily is a northern group barely represented in New Guinea. The beetles are cylindrical and about 5—9 mm long, and usually orange or red and black. *Aetheomorpha* is the only genus recorded.

CRYPTOCEPHALINAE — This is one of the larger subfamilies in temperate countries including Australia, but is much more poorly represented in New Guinea. Fifty-one species have been recorded here (Gressitt 1965b). The species are cylindrical or broadly oblong-ovate, and may be smooth and shiny or grooved or deeply sculptured and tuberculate. Colors are yellow to red, brown, black or metallic blue. Size is 1.25—8 mm. The larvae are somewhat cylindrical and make papery cases. They occur on bark, stems or leaves. Most of the species belong to *Ditropidus, Coenobius, Cryptocephalus* and *Cadmus.*

CHLAMISINAE — Members of this subfamily are similar in shape, size and larval habit to the preceding, but are always black or dark metallic, and are very roughly sculptured and tuberculate. They resemble the droppings of caterpillars of moths and butterflies, and thus are often completely overlooked, especially since they usually sham death when disturbed.

EUMOLPINAE — This subfamily is represented by very many genera and species in New Guinea. The group is better represented here than in most regions, and especially as compared with Australia. The species are for the most part short and broad, chunky to almost spherical, or in some cases 2—3 times as long as broad and oblong or elliptical. Body length is 2—12 mm. Adults may occur in great numbers on trees and shrubs such as *Pipturus, Macaranga, Acalypha, Cordyline, Eurya, Nothofagus,* etc. Some species are metallic and some closely resemble cryptorhynchine weevils and certain small black spiders. Others are brown, reddish or testaceous and more rarely spotted or patterned. The larvae are pale, moderately broad, slightly flattened, and mostly or entirely feed on roots. Among the larger genera are *Stethotes, Deretrichia, Rhyparida, *Rhyparidella, *Phainodina, *Micromolpus, Thyrasia, Cleoporus, Callisina* and *Basilepta.*

CHRYSOMELINAE — This subfamily is moderately represented, but not as abundantly as in Australia. This group includes the largest New Guinea leaf-beetles (*Promechus*) up to 30 mm in length (Pl. 3d; Pl. 9a—c). A number of the species are metallic and very beautiful, with gold, green and blue iridescence (Gressitt & Hart). The larvae of these may be among the largest beetle larvae occurring naked on leaves. Both adults and larvae are fairly long and somewhat flattened. Members of other genera are less elongate, and some are nearly as broad as long. These may be oblong or rounded and convex, some of them resembling ladybird beetles. Among the more conspicuous genera are *Promechus, Paropsis, Stethomela, Phyllocharis* and *Xenolina.*

GALERUCINAE — This is a very large subfamily in New Guinea, but has been little studied here. Species are abundant at all altitudes. They have various shapes and colors, but in general are less compact and less rigid than most other leaf-beetles. They are mostly oblong or oval, somewhat flattened, and broader posteriorly. Some of the larvae are root feeders and some are surface feeders on leaves. This group includes common pests

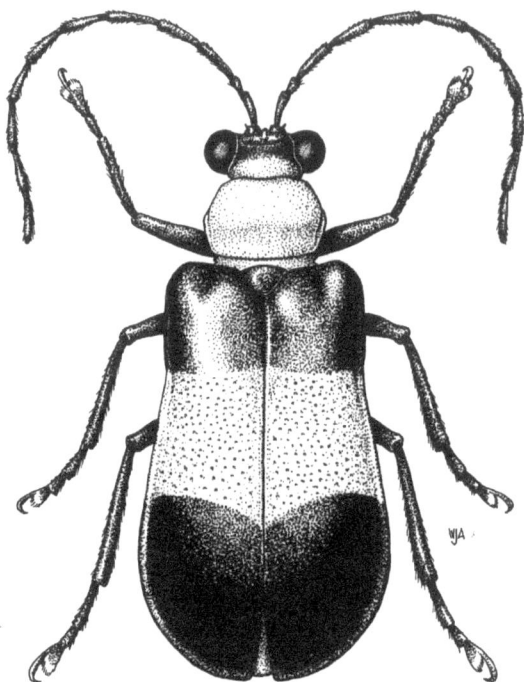

Figure 32 *Aulacophora pallidefasciata* Jacoby
(Chrysomelidae: Galerucinae).

of cucurbits and some other vegetables. Some are called cucumber beetles or squash beetles. Many of these are yellow or orange, banded with black, or entirely orange or mostly black. *Aulacophora* species are also recorded from cassava, *Crotalaria*, etc. Some species are very small, others large and metallic, or sculptured. Length of body is 2–18 mm. Main genera include *Oides, Aulacophora* (Fig. 32), *Prasyptera, Atysa, Sastra, Nicea, Monolepta, Gronovius* (Pl. 9d).

ALTICINAE — This subfamily also is one of the largest in the family and one of the better represented ones in New Guinea. The beetles are very numerous, and usually small to minute (1—6 mm, and rarely 8 or 10 mm in body length). The body is usually short, broader behind, and somewhat flattened. The hind leg, especially hind femur, is stout and the beetles are strong jumpers. Coloration is often brown or black with yellow or red, but may be of other colors, with or without spots or bands. The larvae are mostly pale, fairly slender root feeders. Some species are pests of vegetables and others occur in numbers on common garden or second growth plants. Among principal genera are *Cassena, Xenidea, Crepidodera, Podontia, Ophrida, Psylliodes, Aphthona, Phyllotreta, Longitarsus, Sutrea, Micrepitrix, Amphimeloides, Schenklingia.*

HISPINAE — This subfamily is well represented in New Guinea and nearby islands. The members are usually different in form from those of the rest of the family. Many are elongate-oblong (*Brontispa, Ceratispa*) or elliptical (*Hispodonta*), and flattened above. Such types develop in monocot plants, especially palms, between the main stem and the petiole bases, or in the unfolding fronds (palms, rattans) or between bases of new leaves (*Callistola* Pl. 4a, in *Pandanus, Freycinetia*). Others are leaf-miners in the larval stage and oblong (*Promecotheca* in palms, *Pandanus*), or wedge-shaped as adults (*Gonophora*, in gingers and sedges). Still others, also with leaf-mining larvae, usually in grasses, are quite spiny as adults. The important coconut palm pest *Promecotheca papuana* (see Gressitt 1959b) is in the oblong leaf-mining group. *Brontispa*, also including important coconut palm pests, is in the first group — slender and flat, feeding in the unfolding new fronds. Most of the hispines are tropical, belonging to zones I and II.

CASSIDINAE — This group again is different in form from the rest, being nearly round in outline with the outer margins of prothorax and elytra moderately to greatly expanded. These are the tortoise beetles, many of which are brilliant golden or metallic green-gold in life. They may be weakly to moderately convex above and have short legs largely hidden by the lateral expansions. Some of the tortoise beetles may be confused with ladybird beetles or others. Most members are in zone I. Cassidine larvae are surface leaf-feeders like the adults, and are flattened and attach molted skins to a terminal appendage which may be held above the back and may be covered with excrement. This forms an "umbrella" providing protective camouflage. Among important genera are *Notosacantha, Aspidomorpha* (Pl. 9f), *Laccoptera* and *Cassida.*

Superfamily Curculionoidea
Family ANTHRIBIDAE

This group is related to the weevils and is usually distinguished by a short, broad downward projection of the head, sometimes forming a very short rostrum, or a flat-faced front of the head. The antenna is not elbowed as with many weevils, and is sometimes very long so that such species may be confused with long-horned beetles. Body form is usually oblong and somewhat cylindrical, somewhat slender to quite short. Size varies from about 4–20 or more mm in length, and colors are usually dull browns or reddish, often mottled, but sometimes brighter with sulphury, pale blue-gray, black, white or other colors. The larva is pale, soft-bodied and arched, the head nearrower than the middle of the body. The larvae are mostly wood or stem-borers, but some feed in seeds, fungi, etc. Adults are commonly seen on dead logs or branches. Most are in zones I and II.

Among local genera are *Altipectus, Ancylotrichis* (Pl. 9e), *Apatenia, Apolecta, Araeocorynus, Eucorynus, *Eothaumas, Exillis, Hucus, Idiopus, Litocerus, Mecocerina, Mecotropis, Meganthribus, Phloeobius, Plintheria, Rhaphitropis, Xenocerus, Xylinodes.*

Family ATTELABIDAE

These are short-bodied weevils with a relatively short snout. They are often leaf-rollers, the roll resulting from cuts in a leaf made by the adult female just before laying an egg. The leaf becomes a cylindrical roll around the egg. The larva then feeds in the rolled, but not yet dead, leaf. Members of this group are not abundant in general, and probably occur mostly in zones II and III.

Another small family of weevils is the BELIDAE, which consists of slender cylindrical species with the snout only about as long as the prothorax and projecting forward. This is an Australian group which is more poorly represented in New Guinea.

Family BRENTHIDAE

This family contains many species of elongate, parallel-sided weevils which constitute a conspicuous element in the New Guinea beetle fauna. Some are quite large, being 40–50 mm in length. They are quite narrow, and some are extremely narrow. The head bears a long narrow rostrum which is directed forward and which is of somewhat different shape in the two sexes, the rostrum in the male being frequently widened at the apex. The short antenna is not elbowed as in most weevils, and arises from near the apex of the rostrum in the male, while being attached some distance proximal to the apex in the female. Males are often at least twice as long as the females. The larvae develop in the interior of dead timber, as white grubs with very small thoracic legs. The adults may be found in large numbers on the surface or under the bark of dying or recently dead trees, and at such times numbers of beetles may congregate together. While the beetles are usually uniformly black, brown or fawn in color, some have a patterning, and a large conspicuous species with deep red-brown elytra has vivid yellow parallel streaks.

Among the genera occurring in New Guinea are *Carcinopisthius, Eterozemus, Isomorphus, Eubactrus, Stratiopisthius, Ectocemus, Ithystenus* (Fig. 33), *Miolispa, Orychodes, Pholicodes, *Pseudotaphroderes, Syngenithystenus* and *Tracheloschizus.*

Family APIONIDAE

This is a group of weevils which has not always been separated from the main family, Curculionidae. The members are usually small, often smooth and slender or with elytra rounded at side and wider than the rest of body. The snout is long and slender, usually pointing more or less forward rather than downward. Most members are 10 mm long and are often shiny black or blue, but sometimes partly red or of

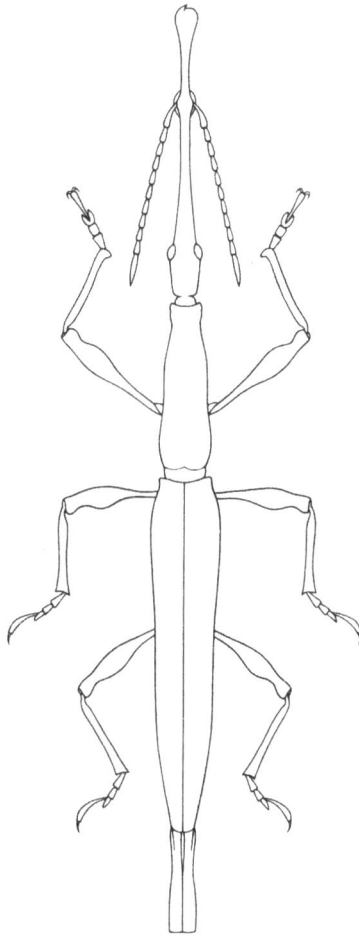

Figure 33 *?Ithystenus* sp. (Brenthidae).

other colors. The group includes some important agricultural pests, such as the sweet-potato weevil, *Cylas formicarius.* The latter name refers to the somewhat ant-like shape of members of this group. Another genus here is *Aphorina* (Pl. 4b).

65

Family CURCULIONIDAE (Weevils; bark beetles)

This group is often referred to as the largest family of animals in the world. Exact limits of the family are not agreed upon, and some families separated earlier have been reduced to subfamilies. Again (see preceding), some former subfamilies are currently regarded as different families. This assemblage form the major portion of the Rhyncophora, often treated as a suborder, but now usually called the superfamily Curculionoidea. The included groups commence with the Anthribidae, above.

Weevils are all associated with plants, and the majority with woody plants. Nearly all parts of plants are fed upon, but many are wood-borers or root-feeders in the larval stage. Others develop in seeds, nuts, fruit, flowers or terminal stems. A few are leaf-miners and many bore under the bark in the cambium. Some weevils have very long lives as adults, and in general weevils are the most hardy insects and quite difficult to kill. Many also have very heavily sclerotized and quite compact bodies.

The most obvious characteristic of many weevils is the long snout or rostrum, which is an elongation of the front portion of the head, with the mandibles located at the end of the snout. Thus the mouth is still a chewing organ as with other beetles, although the snout may be very long and suggestive of a sucking structure. The snout is often fitted into a groove between the legs in repose, but in others it is always prominent and extended forward or downward, and sometimes is as long as the rest of the body.

The general body form of weevils is quite varied, and some are extremely long and slender, and others nearly spherical. Usually, they are elongate-oval or cylindrical, but some are quite rough with tubercles, nodes, ridges or pits, and others have tufts of stout hairs or are clothed with scales. Some weevils have very long legs and may be quite active, even flying rapidly, while others drop and sham death while holding their legs and snout close to the body. Some of the latter are flightless.

The larvae of weevils are usually white or pale creamy, with a brown head, and usually they lack legs. The posterior portion of the body is often stouter than the head and thorax, and the body is often somewhat arched. However, larvae of some are cylindrical or very weakly arched.

The economic importance of weevils is fairly great. Many are associated with agricultural crops, as well as with stored grain, nuts, timber and other plant products. Timber trees are killed by weevils under certain circumstances. This is often as a result of cultural methods such as planting trees in pure stands with resultant upset of normal ecosystem cycles and balances, so that the weevils are favored and their parasites, predators and/or competitors are disadvantaged. Chemical control methods for the pests may result in greater mortality to the natural enemies than to the hardy weevils themselves.

Among major agricultural pests in New Guinea in this family are the serious *Pantorhytes* weevils attacking cacao (I II). These, like most pests in New Guinea, are all native species, each endemic to a limited portion of the area. Very few individual species are widespread, even in a given altitude zone. Some bark beetles are serious pests of *Araucaria*. Among dozens of subfamilies of Curculionidae, the following may be the most important ones in the New Guinea area.

BRACHYDERINAE – This and the following subfamiy together are usually called the "short-nosed weevils" most of which have the snout about as long as broad, and usually broader than in other subfamilies. Members of these two groups are often abundant in individuals, and include a number of species of economic importance. Many of the species of Brachyderinae are leaf-feeders as adults and root- or stem-feeders as larvae. Included are the *Pantorhytes* cacao weevils and their relatives (zones

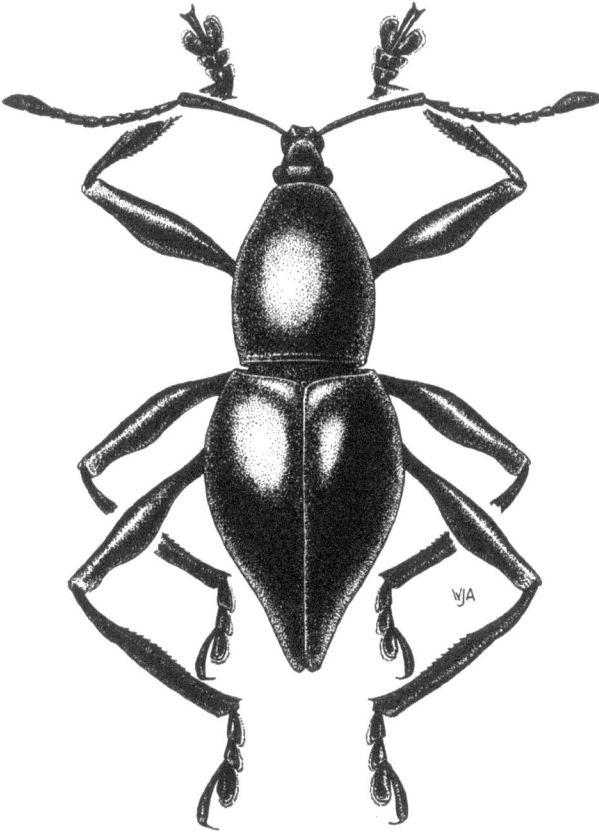

Figure 34 *Behrensiellus glabratus* Pascoe (Curculionidae: Otiorhynchinae).

67

I–III; Pl. 10a) (see Gressitt, 1966b; Stibick, 1977). Members of this group usually have a stout body, with rounded sides to the elytra, and are mostly 6–16 mm in length.

Besides *Pantorhytes*, local genera include *Sphenomorpha* and *Cataphracta*.

OTIORHYNCHINAE – These short-nosed weevils are abundant in numbers of species and individuals and include many agricultural pests, particularly in zones I and II. Many of the species are fairly small, and range from 5 to 12 mm in length. Among common pests attacking various vegetables are species of *Oribius*, such as *O. destructor* (peanuts) and *O. cinereus* (coffee). *Meroleptus cinctor* is another pest of coffee. *Apirocalus cornutus* feeds on a number of kinds of plants, as do species of *Trigonops*. Other genera include *Apocyrtus, Behrensiellus* (Fig. 34), *Coptorhynchus, Elytrocheilus, Hellerhinus* (Pl. 4c; Pl. 9g), and a number of genera in the tribe Celeutethini.

LEPTOPIINAE – This subfamily includes some of the largest weevils in New Guinea, and also some of the most beautiful beetles in the world. The very large ones belong to the genus *Gymnopholus* (zones III, IV; Pl. 9h), the 60-plus species of which only occur above 900 meters altitude on the mainland of the island of New Guinea. Many of the species occurring above 1500 meters altitude have plants growing on their backs (algae, fungi, lichens, liverworts, moss), and there are various minute animals (protozoans, rotifers, nematode worms, plant-feeding mites) living among the plants. (see Gressitt, 1966a, 1971). The genus *Eupholus* (I; Pl. 4d; Pl. 10d), includes the specially beautiful species, some almost entirely covered with brilliant metallic scales, and also rather large in body size, mostly 18–23 mm long. *E. browni* attacks cacao in New Britain. The genus *Rhinoscapha* (I, II; Pl. 10c) also includes fair-sized species, some quite handsome. The adults are leaf-feeders and the larvae are root-feeders, and they are associated with various woody plants, vines or large grasses. Some species may be attracted to light.

MOLYTINAE – This subfamily is a relatively small one, but includes a fairly conspicuous species, *Carbonomassula cobaltina*, which is stout, about 12 mm long and shiny blue-black (Pl. 10e).

RHYNCOPHORINAE (Calendrinae) – This subfamily includes various - and large-sized species, 10–30 mm long, with fairly long snouts. These include the palm weevils, and others attacking bamboos, grasses, bananas and other monocot plants in particular. Of these, the New Guinea sugar-cane weevil (*Rhabdoscelus obscurus*) has been introduced to certain other sugar-cane growing regions. Species of *Sphenophorus* also attack sugar cane. Among the palm weevils are *Rhyncophorus* species (Pl. 10f) which are quite large, and *Sparganobasis* and others which are much smaller. Other genera include *Barystethus, Ciatheles* and *Aporosphenus*.

HYLOBIINAE – This subfamily includes various wood-boring species,

68

among them the very large *Vanapa oberthueri* (Pl. 10b) which attacks *Araucaria* trees. The weevils are shiny black with rows of pits and the snout is extremely long. *Orthorhinus* is another common member.

ALCIDENINAE – This group includes slender, more or less cylindrical, species with fairly long snouts. *Alcides* includes some common species, some of which bore in large herbs and have fairly cylindrical larvae.

LIXINAE – This subfamily includes slender cylindrical weevils which often are covered with a white powdery wax. They are mostly stem-borers in herbs or semi-woody plants and the larvae are more or less cylindrical. The genus *Lixus* includes some agricultural pests.

CRYPTORHYNCHINAE – This is one of the largest subfamilies in New Guinea. Members are from 2 to 20 mm or so in length, and many have tubercles, nodes, ridges, etc. Some are flightless. All have a groove between the legs for the reception of the rostrum. Body form is often oval and the body is usually deep and narrowed posteriorly. When disturbed the weevils often sham death and fall like seeds. Many of the species feed in seeds, nuts and fruit, and others bore in stems of living plants, or in logs. Among the genera are *Conomalthus, Eudyasmus, Ittostira, Meroleptus, Perissops, Ptolycus, Poropterus, Odosyllis, Ectatocyba, Disopirhinus* (Pl. 10g).

SIPALINAE – This is a small subfamily of large weevils which have a long curved snout and are generally brown with rows of darker spots. The larvae bore in logs. Common species belong to *Sipalis*.

ZYGOPINAE – Members of this subfamily have very long legs and run rapidly on logs and take flight very quickly. Some of them have long fringes of fine hairs on each side of each tibia. The weevils are oval, with body from 8–12 mm long as a rule, and are often pale gray or black and white. *Arachnopus* (I; Pl. 10h) is a large genus in this group and others include *Mecopus*.

BARIDINAE – This group includes mostly medium-sized weevils which are often brown, black or speckled, more or less cylindrical, and clothed with dull scales. Most of the species are wood-borers as larvae. New Guinea genera include **Aulacobaris* and *Pseudochlus*.

ERIRHININAE – Members of this subfamily are mostly medium-small, stout, and brown or dull in color. The snout is moderately long and round in cross-section. Included is *Amorphoidea*.

BALANININAE – This subfamily includes fairly broad weevils, narrowed behind, with the snout slender and as long, or longer, than the body, and the legs fairly long and slender. Genera include *Balaninus* and *Carponinus*. They feed in nuts.

SCOLYTINAE – This subfamily appears as the family Scolytidae in most older texts. These are the true bark-beetles, although the following subfamily, now also considered within the family of weevils, is also associated with the under-bark habitat, as are many other weevils and members of some other beetle families, such as the longicorn beetles. Scolytines are very numerous in species and are often abundant as individuals in a single tree or log. They are mostly quite small; 2–8 mm in length. Many species are somewhat slender and cylindrical, but others are short and stout. Color is usually reddish brown to blackish. The antenna is short and strongly clubbed. The prothorax may have series of small tubercles or ridges, as may the elytra, especially the posterior declivity. The larvae are pale, soft and wider behind. They bore tunnels under bark. The female may lay eggs along a tunnel it makes and the larvae bore outward from this tunnel, often at right angles, or radiating, to form a distinctive pattern both on the surface of the trunk or log, and on the inner surface of the bark. Since all the feeding is in the cambial area, great damage is done to the trees, and these frequently die as a result. The death of the tree may be partly the result of the action of fungi spread by the parent bark beetles. Some species, called 'ambrosia beetles,' cultivate certain fungi upon which the larvae feed. The adults of these species have special cavities on their bodies in which the fungal spores are transported from one burrow to another and the behavior of the beetles assists in this dissemination. With such habits the species can be much more harmful to the trees than would be the case if they were only boring through the wood. Schedl (1968 etc) has described many species. New Guinea genera include *Allarthrum, Hylesinus, Hylurdrectonus, Ozopemon, Poecilips, Scolytus, Xyleborus* (Fig. 35).

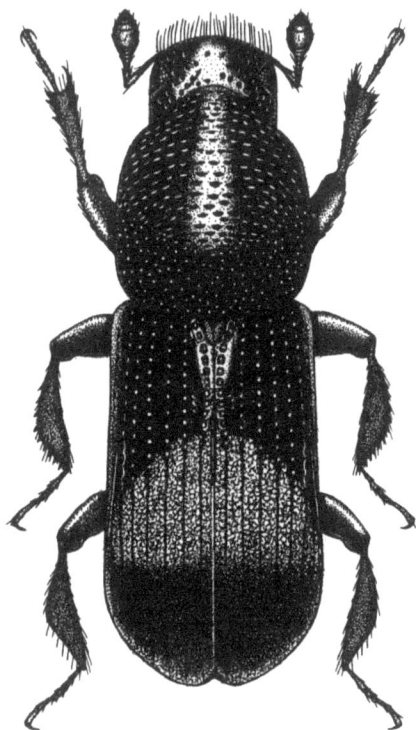

Figure 35 *Xyleborus sordicavola* Schedl (Curculionidae: Scolytinae).

PLATYPODINAE – This group is closely related to the Scolytinae and likewise was formerly considered as a separate family. The platypodines are more elongate than most scolytines, and are a bit less cylindrical as a rule, being slightly flattened above. They are parallel-sided and often trun-

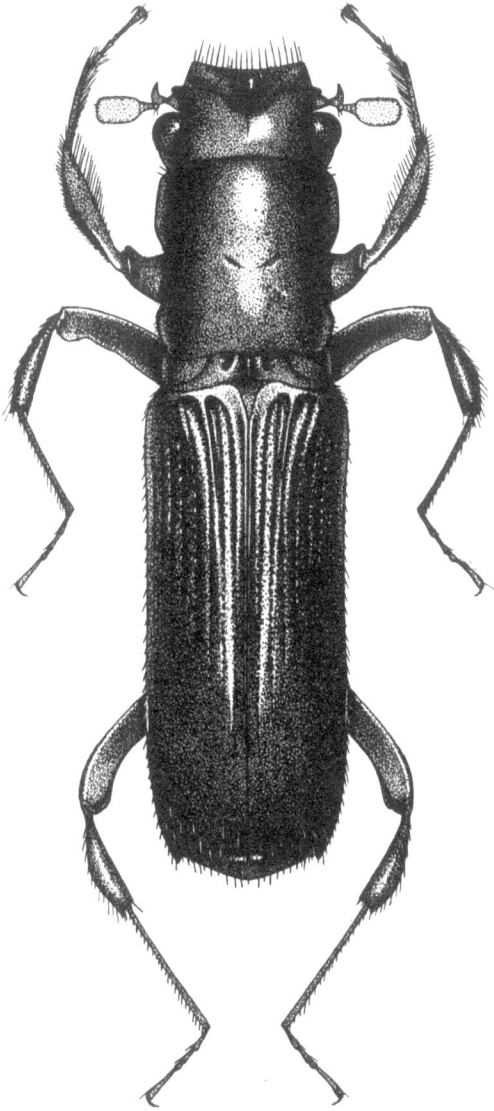

Figure 36 *Crossotarsus bironeanus* Schedl
(Curculionidae: Platypodinae).

cate and oblong. Most species range about 4–10 mm in body length. The antenna is likewise short and the tarsi are long and slender. Most species feed under bark, or deeper in trunks and logs. The larvae are similar to those of scolytines. Local genera include *Crossotarsus* (Fig. 36), *Diapus, Platypus, Spathidicerus.*

COLLECTION AND PRESERVATION OF SPECIMENS

Collecting and studying beetles is not merely a museum undertaking, but may be a stimulating and satisfying hobby. There is a real place for collecting beetles, partly to satisfy the inquisitive collecting instincts of people, but also to stimulate and inspire an interest in natural history, and natural phenomena. In a country such as Papua New Guinea, identified and well preserved beetles collections will definitely provide vital material for the study and understanding of the beetle fauna of the country. Collections may be specialized to include single groups or families, and collectors may learn to investigate special habitats or use special techniques depending on the group of beetles of their specialty. This may result in unequalled collections of the group concerned, with many new species. In general, virtually every situation from a backyard garden to the msot remote area in the country is worthy of study. Some regions are particularly interesting and important, having been neglected. Certain habitats are very rich in species and individuals. For instance, recently felled bush, being prepared for a garden, is richly productive of several families of beetles.

Beetles are easier to collect and keep than most insects because of their relatively strong exoskeletons. They are much less fragile than most other insects. Beetles may be killed for specimens by cyanide, ethyl acetate (acetic ether), or alcohol, or even with boiling water. The chemicals are all more or less dangerous, especially cyanide, and should be handled carefully, and used in tightly corked tubes or jars. Cotton-wool and cardboard, or plaster of paris, is used to confine the chemical. Ethyl acetate should be used with glass jars or tubes, as it can dissolve plastic. It has the advantage of being less dangerous than cyanide and in leaving specimens relaxed and temporarily mold-resistant, thus reducing breakage and deterioration. Larvae may be preserved in a fluid consisting of surgical spirit (alcohol) – 7 parts; glacial acetic acid – 1 part; and water – 2 parts; or some other special formula.

The beetles should be pinned through the base of the right elytron, with standard insect pins. Use sizes 4 to 6 for very large specimens, 3 for medium-large ones, 2 for medium-small. Minute specimens may be glued at the right side of the body on the end of a slender point (narrow triangle) of stiff white paper. The end of the point is bent down to fit the side of the thorax and the insect pin goes through the wide end of the paper point (about 8 mm long). A very small amount of glue should be used in order to avoid hiding body structures. Another method is to glue the beetle on an oblong card, the pin going through the base of the card. However, this system has the disadvantage of hiding ventral structures often used in identification. Still another method is to use "Minuten" needles into the right side of the beetle, horizontally, with both large and small pins through a small block of pith.

One or more labels in India ink on rag paper should indicate country, province, town or defined locality, with latitude, longitude, altitude, date, collector and host and/or habitat. The labels are placed on the same pin, beneath the body of the beetle, spaced so they can be read without removal, facing to left and in line with the body of the beetle. There are small hand presses available for printing labels. Another method is to type required labels in columns, with 18 or so letters per line, with say 3 to 5 lines per label, and to have the labels printed by photo-offset.

To prevent museum beetles or other pests from eating the dried and pinned specimens, the beetles should be kept in a tight cork-lined box or in a tray inside a tightly enclosed case with some repellent like naphthalene or paradichlorobenzene. The latter gives off a toxic gas, so should be used sparingly and only in tight containers for fumigation, or temporarily if naphthalene is unavailable. Some people are allergic to naphthalene. An effective alternative is beechwood creosote, a liquid. It can be kept in a small tube with cotton wick or just a drop put on inside of cover of box. Only minute amounts should be used, as the gas is harmful and corrosive. These chemicals, especially the creosote, will also prevent mold, but the specimens must be well dried also, and should be examined from time to time. If mold commences, the beetles can be cleaned with ethyl acetate, using a very fine brush and taking great care with antennae, palpi, legs and hairs. If the specimen is brittle, it may be relaxed over night in a moisture chamber with a bit of creosote, or wet with ethyl acetate for a while. After cleaning they should be more thoroughly dried and kept dry. A tight cabinet with an electric bulb installed in the base, or a closet or small room with a dehumidifier, should prevent molding.

REFERENCES

The following references include some of the major works on New Guinea beetles, as well as the items cited in the text, plus some dealing with neighboring areas which may aid the student considerably.

Arrow, G. J. 1951. Horned beetles, a study of the fantastic in nature. 154 p., W. Junk, Den Haag. [Scarabaeidae, etc.]

Baly, J. S. 1865. Phytophaga Malayana. *Trans. Ent. Soc. London* ser. 3, 4 (1): 1–300, 5 pls. [Chrysomelidae]

Böving, A. G. & F. C. Craighead. 1931. An illustrated synopsis of the principal larval forms of the order of Coleoptera. *Entomologica Americana* 11 (1–4): 1–351, 125 pls.

Britton, E. B. 1970. Coleoptera (beetles). In: The insects of Australia, p. 495–621. Melbourne Univ. Press.

Crowson, R. A. 1967. The natural classification of the families of Coleoptera. (Reprint, E. W. Classey Ltd, Middlesex) 214 p.

Darlington, P. J. 1952. The carabid beetles of New Guinea. Part 2. The Agonini. *Bull. Mus. Comp. Zool.* 107 (3): 89–252, 4 pl.

1962. The carabid beetles of New Guinea. Part 1. Cicindelinae, Carabinae, Harpalinae through Pterostichini. *Bull. Mus. Comp. Zool.* 126 (3): 321–564, 4 pl.

1968. The carabid beetles of New Guinea. Part 3. Harpalinae (continued): Perigonini to Pseudomorphini. *Bull. Mus. Comp. Zool.* 137 (1): 1–253.

1971. The carabid beetles of New Guinea. Part 4. General considerations; analysis and history of fauna; taxonomic supplement. *Bull. Mus. Comp. Zool.* 142 (2): 129–337.

Endrödi, S. 1971. Monographie der Dynastinae. 2. Tribus: Oryctoderini (207–241). 4. Tribus: Pentodontini (papuanische und pazifische Inselwelt) (243–320) (Coleoptera: Lamellicornia: Melolonthidae). *Pacific Insects* 13 (2). [Scarabaeidae]

Fairmaire, L. 1883. Essai sur les Coléoptères de l'Archipel de la Nouvelle-Bretagne. *Ann. Soc. Ent. Belg.* 27 (2): 1–58.

Fauvel, C. A. A. 1878. Les staphylinides des Moluques et de la Nouvelle Guinée. *Ann. Mus. Civ. Genova* 12: 171–315, 2 pl.

Fleutiaux, E. 1896. Eucnémides Austro-Malais du Musée Civique de Gênes. *Ann. Mus. Civ. Genova* 36: 555–606.

Froggatt, J. L. & B. A. O'Connor. 1941. Insects associated with the coconut palm. Pt. 2. *New Guinea Agric. Gaz.* 7 (2): 125–133.

Gebien, H. 1920. Coleoptera, Tenebrionidae. Résultats de l'expedition scientifique Néerlandaise à la Nouvelle-Guinée .... *Nova Guinea* 13 (3): 213–500, 3 pl.

Gestro, R. 1875. Descrizione di un nuovo genere e di alcuni nuove species di Coleotteri papuani. *Ann. Mus. Civ. Genova* 7: 993–1027.

1876. Enumerazione dei Longicorni della tribu dei Tmesisternini, raccolti nella regione Austro-Malese dai signori O. Beccari, L. M. d'Albertis e A. A. Bruijn. *Ann. Mus. Civ. Genova* 9: 139–174.

Gilmour, E. F. 1959. Revision of the genus *Rosenbergia* Ritsema (Coleoptera, Cerambycidae, Lamiinae, Batocerini). Idea 12: 21–51. Also 1960, *Idea* 13: 1–34; 1966, *Reichenbachia* 6: 245–261.

Gravely, F. H. 1914. An account of Oriental Passalidae (Coleoptera) based primarily on the collection in the Indian Museum. *Mem. Ind. Mus.* 3 (4): 177–353, 3 pl.

Gressitt, J. L. 1959a. Longicorn beetles from New Guinea. 1 (Cerambycidae). *Pacific Insects* 1: 59–171.

1959b. The coconut leaf-mining beetle *Promecotheca Papuana. Papua New Guinea Agric. J.* 12: 119–147, 8 pl.

1961. Problems in the zoogeography of Pacific and Antarctic insects. *Pacific Insects Monograph* 2: 1–94.

1963. Hispine beetles (Chrysomelidae) from New Guinea. *Pacific Insects* 5: 591–714.

1964. Economic chrysomelid beetles from New Guinea, with new species. *Papua New Guinea Agric. J.* 16: 105–116.

1965. Chrysomelid beetles from the Papuan Subreion, 1 (Sagrinae, Zeugophorinae, Criocerinae). *Pacific Insects* 7: 131–189. (Parts 2–8, 1965–1974, *Pacific Insects* vols. 7–16). (8: Gressitt & Hart)

1966a. Epizoic symbiosis: the Papuan weevil genus *Gymnopholus* (Leptopiinae) symbiotic with cryptogamic plants, oribatid mites, rotifers and nematodes. *Pacific Insects* 8: 221–280.

1966b. The weevil genus *Pantorhytes* (Coleoptera), involving cacao pests and epizoic symbiosis with cryptogamic plants and microfauna. *Pacific Insects* 8: 915–965.

1974. Insect biogeography. *Ann. Rev. Entom.* 19: 293–321.

Gressitt, J. L. & J. J. H. Szent-Ivany. 1968. Bibliography of New Guinea entomology. *Pacific Insects Monograph* 18: 1–674.

Grouvelle, A. H. 1882. Cucujides nouveaux du Musée Civique de Gênes. *Ann. Mus. Civ. Genova* 18: 275–296.

76

Heller, K. M. 1895. Erster Beitrag zur Papuanischen Käferfauna. *Abh. Ber. Zool. Anthr. -Ethn. Mus. Dresden* 5 (16): 1–17. Continuation in same serial, 1897, 6 (11); 1901, 10 (2); 1908, 12 (1); 1910, 13 (3). [Misc. Coleoptera]

— 1914. Coleoptera. Résultats de l'expéd. néerlandaise à la Nouvelle Guinée. *Nova Guinea* 9 (5): 615–667, 2 pl.

— 1926. Coleoptera: Curculionidae. Résultats des expéditions scientifiques à la Nouvelle Guinée. *Nova Guinea* 15: 275–291.

— 1934. Käfer aus dem Bismarck und Salomo-Archipel. *Verh. Naturf. Ges. Basel* 45: 1–34, 1 pl. [Misc. Coleoptera]

— 1935. Coleoptera, Curculionidae: Résultats des expéditions scientifiques à la Nouvelle Guinée. *Nova Guinea* 17 (2): 155–202, 1 pl.

Hincks, W. D. 1937. Passalidae (Col.) from Papua collected by Miss L. E. Cheesman. *Nova Guinea*, n. s. 1: 112–124.

Hinton, H. E. 1945. Monograph of the beetles associated with stored products. Brit. Mus. (Nat. Hist.) 1: 1–443.

Jacoby, M. 1884. Descriptions of new genera and species of phytophagous Coleoptera from the Indo-Malayan and Austro-Malayan subregions, contained in the Genoa Civic Museum, Pt. 1. *Ann. Mus. Civ. Genova* 20: 188–233. (pt 2, 1885, 22: 20–76). (pt 3, 1886, 24: 41–128). [Chrysomelidae]

Jordan, K. 1924. New Anthribidae. *Novit. Zool.* 31: 231–255.

Kalshoven, L. G. E. 1961. Habits and host-associations of Indomalayan Rhynchophorinae (Coleoptera, Curculionidae). *Beaufortia* 9: 49–73.

Kaszab, Z. 1939. Tenebrioniden aus Neu Guinea. Nova Guinea, n.s. 3: 185–267.

— 1977. Die Tenebrioniden des Papuanischen Gebietes I. Strongylini. *Pacific Insects Monogr.* 33: 1–227.

Kerremans, C. 1904–1913. Monographie des buprestides. 8vo Bruxelles, 6 vols.

Kleine, R. 1926a. Coleoptera, Lycidae. Résultats de l'expédition scientifique néerlandaise à la Nouvelle Guinée. *Nova Guinea* 15: 91–195.

— 1926b. Die Brenthiden des papuanischen Gebietes. Résultats de l'expédition scientifique néerlandaise à la Nouvelle Guinée. *Nova Guinea* 15: 214–274.

Laboissiere, V. 1932. Coleoptera: Galerucinae. Résultats scientifiques du voyage aux Indes orientales Néerlandaises de LL. AA. RR. le Prince et la Princesse Léopold de Belgique. *Mem Mus. R. Hist. Nat. Belg.* hors ser. 4 (4): 145–184, 2 pl. [Chrysomelidae]

Lacordaire, J. T. 1866. Genera des Coléoptères, ou exposé méthodique et critique de tous les genres proposés jusqu'ici dans cet ordre d'insectes. 12 vols, 8vo, Paris.

Lisle, M. O. de. 1967. Note sur quelques Coleoptera Lucanidae nouveaux ou peu connus. *Rev. Suisse Zool.* 74: 521–544.

Lucassen, F. T. Valck. 1961. Monographie du genre *Lomaptera* Gory & Percheron (Coleoptera, Cetonidae). Amst. Nederl. Ent. Vereen. 299 p.

Marshall, G. A. K. 1956. The Otiorrhynchine Curculionidae of the tribe Celeuthetini (Col.). British Museum (Nat. Hist.), Lond., 134 p, 1 pl.

1957. Some injurious Curculionidae (Col.) from New Guinea. Bull. Ent. Res. 48: 1–7.

1959. Curculionid genus Gymnopholus (Coleoptera). *Occ. Pap. B. P. Bishop Mus.* 22 (7): 71–81.

Ochs, G. H. A. 1955. Die Gyriniden-Fauna von Neu-Guinea nach dem derzeitigen Stand unserer Kenntnisse (Coleoptera, Gyrinidae). Results of the Archbold Expeditions. *Nova Guinea*, n.s. 6 (1): 87–154.

Pascoe, F. P. 1864–1869. Longicornia Malayana; or a descriptive catalogue of the species of the three longicorn families – Lamiidae, Cerambycidae and Prionidae, collected by Mr. A. R. Wallace in the Malayan Archipelago; summary & locality notes by A. R. Wallace. *Trans. Ent. Soc. Lond.* ser. 3, 3: 1–712, 24 pl. [Cerambycidae]

1885. List of the Curculionidae of the Malay Archipelago collected by Dr. Edoardo Beccari, L. M. D'Albertis and others. *Ann. Mus. Civ. Genova* 22: 201–332, 3 pl.

Samuelson, G. A. 1967. Alticinae of the Solomon Islands (Coleoptera: Chrysomelidae). *Pacific Insects* 9 (1): 139–174.

Schedl, K. E. 1968. On some Scolytidae and Platypodidae of economic importance from the Territory of Papua and New Guinea. *Pacific Insects* 10 (2): 261–270.

Schenkling, S. et al. 1910–. Coleopterorum Catalogus & Supplements. W. Junk, Berlin & Den Haag.

Simon Thomas, R. T. 1964. Some aspects of life history, genetics, distribution and taxonomy of *Aspidomorpha adhaerens* (Weber, 1801) (Cassidinae, Coleoptera). *Tijds. Ent.* 107 (4): 167–264.

Stibick, J. N. L. 1978. The cacao weevils of New Guinea. *Pantorhytes.* Division A, new species, synonymy and key (Coleoptera: Curculionidae). *Pacific Insects* 19 (1):

Szent-Ivany, J. J. H. 1959. Host plant and distribution records of some insects in New Guinea. *Pacific Insects* 1 (4): 423–429.

1961. Insect pests of *Theobroma cacao* in the Territory of Papua and New Guinea. *Papua New Guinea Agric. J.* 13 (4): 127–147, 3 pl.

1965. Notes on the vertical distribution of some beetles in New Guinea with new locality data and host plant records of some high altitude species. *Papua New Guinea Sci. Soc. Trans.* 6: 20–36.

Szent-Ivany, J. J. H. & A. Catley. 1960. Host plant and distribution records of some insects in New Guinea and adjacent islands. *Pacific Insects* 2 (3): 255–261.

Van Zwaluwenburg, R. H. 1963. Some Elateridae from the Papuan region (Coleoptera). *Nova Guinea, Zool.* 16: 303–346.

Voss, E. 1956–1960. Die von Biro auf Neu-Guinea aufgefundenen Rüsselkäfer (Col.). I-III. *Ann. Hist. Nat. Mus. Nat. Hung.* n.s. 7: 121–142; 9 (50): 209–222; 52: 313–346. [Curculionidae]

Weise, J. 1917. Chrysomeliden und Coccinelliden aus Nord-Neu-Guinea gesammelt von Dr. P. N. van Kampen und K. Gjellerup, in den Jahren 1910 und 1911. *Tijds. Ent.* 60: 192–224.

Wittmer, W. 1964. Neue Malacodermata aus Neu Guinea. *Nova Guinea, Zool.* 30: 115–137. [Cantharidae]

1969. Zur Kenntnis der indo-malaiischen Silini unter besonderer Berücksichtigung der Fauna von Neuguinea (Col.: Cantharidae). *Pacific Insects* 11 (2): 217–454.

1971. Zur Kenntnis der Cantharidae (Col.) Neugeuineas und angrenzender Gebiete. *Pacific Insects* 13 (3–4): 545–574.

GLOSSARY

AEDEAGUS — Sclerotized portion of male reproductive organ.

ANALOGOUS — Similar, as in structure and/or function, but not necessarily of same evolutionary origin (latter = homologous).

ANTENNA — Paired and segmented feeler inserted more or less on top of front of head, or on side before eye.

APPENDICULATE — Bearing appendages, as in having secondary tooth near base of each tarsal claw.

BIOMASS — The total weight of living and standing material in a given area, including all animals and plants.

BIOTA — The flora plus the fauna; thus all living things (kinds) in an area.

CATERPILLAR — The immature growing form of a moth or butterfly.

CERCI — A pair of appendages at end of abdomen; lacking in most beetles.

COMPOUND EYE — Paired large eye on side of head, with many minute facets.

DIURNAL — Active in the daytime, as opposed to nocturnal.

ECOSYSTEM — Sum total of interacting species and factors in a limited habitat, say of one vegetation type with its animal inhabitants as well.

ELYTRON — (plural, elytra) — The hardened wing-covers (fore wings).

EMPODIUM — A pad or seta protruding between the tarsal claws.

ENDEMIC — Restricted to a given area, such as island of New Guinea, or a more limited area, such as the Huon Peninsula, or a single mountain range (applied to species, genus, or even a higher category).

EXOSKELETON — Sclerotized outer wall or shell of body and appendages.

FALSE 4-SEGMENTED (of tarsus) — With 4th segment minute and 3rd lobed.

FAMILY — A large (higher) category usually including many genera and species, and often a number of subfamilies and tribes, but smaller than superfamily and order.

FILIFORM — Slender, with segments uniformly narrow.

FLABELLATE — Fan-shaped, as with antenna with projections on each side of each segment or most segments.

FOOD-CHAIN — Series of steps in an ecosystem representing the feeding of one species on another, including an insect eating a plant, for instance, with another insect feeding on the first insect, and parasites, birds, etc feeding on each of these, and so on.

FRONT — Plate or area on the front of the head, usually between antennal insertions and often below or between eyes.

GENICULATE — Elbowed, as in antenna with 1st segment very long and rest at angle to it.

GENUS (pl., genera) — Taxonomic unit including related species. The first of the usual two parts of the scientific name of a species.

GLABROUS — Without hairs; naked.

GULAR SUTURE — Groove on underside of head, single (median) or double.

HOMOLOGY (adj, homologous) — Of common origin, as structures of similar or different appearance but of same ancestral (evolutionary) origin.

INDIGENOUS — Native to an area (not introduced by man); but may also occur elesewhere, and if so is not endemic.

INSTAR — Individual of a given stage (stadium) of a larva (between molts).

LABIUM — Equivalent of lower lip; consisting of partly fused paired 3rd jaws; bearing a pair of palpi.

LARVA (pl., larvae) — Immature growing stage of insects, between egg and pupa (same as caterpillar, but latter term is applied only to moth and butterfly larvae).

MAXILLA (pl., maxillae) — 2nd pair of jaws, used in manipulating food; each bearing a palpus.

METEPISTERNUM — Plate on side of hind thorax, often oblong and parallel to and close to lower border of elytron.

MOLT — The shedding of old exoskeleton with formation of a new, larger one.

MUTUALISM — A relationship between two species whereby both benefit.

NEW GUINEA — Here referring to the entire island of New Guinea, including Papua and Irian Jaya.

OCELLUS (pl., ocelli) — Small simple eye (often 2 or 3) between or in front of compound eyes; lacking in most beetles.

ORDER – Major group within the insects (or birds, etc), as all the beetles together; usually divided into suborders, superfamilies and families.

PALPUS (pl., palpi) – Small segmented taster (rarely longer than antenna); a pair each on maxillae and labium, under mandibles.

PAPUAN – Referring to Papuan Subregion, including Moluccas, Irian Jaya, Papua New Guinea and Solomons.

PARASITE – Organism living at expense of another, in or on its body.

PECTINATE – Comb-like, as with long extentions on one side of each antennal segment.

PUPA – Resting stage of life cycle for change (metamorphosis) from larva to adult.

PYGIDIUM – Dorsal surface of last visible abdominal segment, often hidden by elytra.

PREDACEOUS – Feeding upon live insects or other animals.

SCAVENGING – Feeding upon dead animal or plant material.

SCLEROTIZED – Hardened structures such as exoskeleton of insects containing chitin.

SERRATE – Toothed or saw-like, as with some beetle antennae.

SPECIES – The unit of kinds of living things; members of a breeding population.

SPERMATHECA – Organ in female in which male sperm are stored.

STRIATE – With rows of grooves or rows of punctures, as on elytra.

STRIDULATION – Sound production by rubbing structures, as legs and elytron, or pro- and mesothorax.

SUBFAMILY – Subdivision of a family; grouping of related tribes or genera.

SUBSPECIES – A geographical or biological race of a species, isolated in some way from other populations.

SUPERFAMILY – A group of families with some characters in common.

SYMBIOTIC – Practicing symbiosis, a form of mutualism involving dependence upon another species, or two each dependent upon the other.

TARSAL CLAWS – Usually paired claws at end of each tarsus (foot).

TENERAL – Condition of newly emerged adult before sclerotization and pigmentation are complete.

THORAX – The middle portion of an insect body. The fore part (prothorax) usually enlarged and distinct in beetles; the hind part (meso- and metathorax) hidden from above by elytra.

TRUNCATE – Transversely or obliquely cut off behind, as with some elytra.

INDEX

References to species illustrated in the plates are given in bold face, as **la**.

85